Drean

T0074516

'If you want a glimpse into how our desires are manufactured under our
new techno-feudal order, and how love can prove our last defence, read
Alfie Bown's *Dream Lovers*.'
—Yanis Varoufakis

'Alfie Bown is one of the smartest writers around. *Dream Lovers* is an
exciting, astute analysis of how our capacity for desire and reverie has
been slotted into the grooves of digital capitalism, and made to work
for profit – from porn to Pokémon.'
—Richard Seymour, author of *The Twittering Machine*

'A brisk and engaging examination of the gamification of love,
a sharp analysis of contemporary technopolitics and of the synthetic
quality of desire.'
—Helen Hester, Professor of Gender, Technology and Cultural
Politics, University of West London, and co-author of *After Work: The
Fight for Free Time*

'Informative, entertaining and devastating all at once. Extremely
well-argued and persuasive, it clearly shows how the technologies that
promise to help us find love, or even just sex, are also perfectly designed
to exploit us at our most vulnerable.'
—Anouchka Grose, writer and psychoanalyst, Centre for Freudian
Analysis and Research

Digital Barricades:
Interventions in Digital Culture and Politics

Series editors:
Professor Jodi Dean, Hobart and William Smith Colleges
Dr Joss Hands, Newcastle University
Professor Tim Jordan, University College London

Also available:

Dream Lovers
The Gamification
of Relationships

Alfie Bown

First published 2022 by Pluto Press
New Wing, Somerset House, Strand, London WC2R 1LA

www.plutobooks.com

British Library Cataloguing in Publication Data
A catalogue record for this book is available from the British Library

ISBN 978 0 7453 4488 1 Hardback
ISBN 978 0 7453 4487 4 Paperback
ISBN 978 0 7453 4491 1 PDF
ISBN 978 0 7453 4489 8 EPUB

This book is printed on paper suitable for recycling and made from fully
managed and sustained forest sources. Logging, pulping and manufacturing
processes are expected to conform to the environmental standards of the
country of origin.

Typeset by Stanford DTP Services, Northampton, England

Simultaneously printed in the United Kingdom and United States of America

Contents

Series Preface

Crisis and conflict open up opportunities for liberation. In the early twenty-first century, these moments are marked by struggles enacted over and across the boundaries of the virtual, the digital, the actual, and the real. Digital cultures and politics connect people even as they simultaneously place them under surveillance and allow their lives to be mined for advertising. This series aims to intervene in such cultural and political conjunctures. It features critical explorations of the new terrains and practices of resistance, producing critical and informed explorations of the possibilities for revolt and liberation.

Emerging research on digital cultures and politics investigates the effects of the widespread digitisation of increasing numbers of cultural objects, the new channels of communication swirling around us and the changing means of producing, remixing and distributing digital objects. This research tends to oscillate between agendas of hope, that make remarkable claims for increased participation, and agendas of fear, that assume expanded repression and commodification. To avoid the opposites of hope and fear, the books in this series aggregate around the idea of the barricade. As sources of enclosure as well as defences for liberated space, barricades are erected where struggles are fierce and the stakes are high. They are necessarily partisan divides, different politicizations and deployments of a common surface. In this sense, new media objects, their networked circuits and settings, as well as their material, informational, and biological carriers all act as digital barricades.

Jodi Dean, Joss Hands and Tim Jordan

Introduction: The Grindr Saga

Dreams spring from reality and are realized in it.
– Ivan Chtcheglov (1953)

Wandering into the intoxicating world of bizarre and alluring pleasures that is the Tencent App Store, the world's largest digital marketplace, you'll probably experience an initial sense of confusion. Bombarded with gauche images, garish pop art text and bold promises of riches, orgasms and other delights, it would be hard not to feel at least somewhat overwhelmed. Over time, though, you get accustomed to this dream world of hallucinogenic commodities, and you might even find a little corner of it in which you can find some relaxation, distraction – or perhaps even pleasure – of your own.

Perhaps you navigate past the invitations of scantily-clad casino hosts, specialised dating sites and personally tailored pornography (now under greater censorship there), and bypass the artificial intelligence (AI) chatbots, virtual reality (VR) relationship simulators and windows of livestreamers vying for your click and find yourself in the world of 希望：再次遇見你, or *Hope: To See You Again*, an animated videogame made by one of the larger games companies in China. In that imaginary world you exist as an avatar, get attached to a Tamagotchi-like robopet and set off on the long adventure of a second virtual life, ending in your marriage to an Anime-style fembot with whom you see out your days in blissful pastoral serenity. It might all be a bit strange, but it's familiar enough when it's safely confined to the virtual world of videogames and fantasy that we have become familiar with vacationing in through our screens.

Until it isn't. Between 2016 and 2018 Beijing Kunlun Tech, the company who own *Hope: To See You Again* and many of the world's most popular mobile and online games completed a takeover of Grindr. Grindr, the world's largest gay dating app, had been a US-based company since its inception in 2009 and one of the few that had remained independent. Until then it had escaped the clutches of IAC, the New York holding company whose offices are housed in a glass blob worthy of science fiction that gazes across the water from Manhattan to New Jersey. Until 2020 IAC owned Match Group, the umbrella over almost all the major US dating sites, from PlentyofFish and OKCupid to Hinge and Tinder, all discussed in this book. Dating sites were grouped with each other, sharing data, algorithms, profits, and so on. Then, the love industry seemed like something of a closed circle, dominated by one conglomerate and operating according to its own logic. Now, the gap between games and love, and between simulation and reality, is closing.

The Grindr takeover embodies this trend. Of course, one significance of the takeover is the increasing globalisation of platforms, of the corporate interest in and use of their data and specifically of China's increasingly powerful role in these patterns. Access to the data that comes from dating sites is empowering. It is intimate data. In carrying out our sexual lives and our love lives at least partially and often entirely online, we create powerful data connected to our most intimate and ultimate desires. This data has great value, both as the basis for targeted advertising campaigns but also – as we shall see in this book – as the basis for a wider and more complete social reorganisation. By knowing and changing what we desire, platforms and tech companies are empowered to manufacture the future of social life, of sex and even of love. In 2020, under direct pressure from then President Donald Trump, Kunlun were forced to sell Grindr back to the US precisely because of fears about the handover of power to China that this powerful data would facilitate.

The Grindr saga is almost humorous given that it ended with a misogynist Trump getting his gay dating site back from the clutches of a Chinese government who themselves refuse to acknowledge LBGTQ+ rights. Perhaps the moral of that story is that money talks and that cultural ideologies can be put aside fairly easily if need be. But there is an even more significant lesson from the Kunlun Grindr connection: the nascent gamification of love. Though Grindr might be a perfect early example of this gamification, on a wider scale this new gamification of love is a process dominated by heteronormative and conformist politics.

As such, a lot of the material in this book – from sexbots to smart condoms to videogames and VR pornography – can be seen as part of this pattern that is dominated by capitalism, heteronormativity and masculinity. On occasion, radical 'alternative' examples of experimental technologies that have attempted to cut against the grain of these trends are mentioned, but on the whole there is less said here about race, trans, neurodivergent and disability politics than there might be. The project of the book is to identify the normative and dominant trends that have been reshaping the world of love in the service of heteronormativity, capitalism and inequality, so the majority of examples discussed here reflect this pattern in order to criticise and begin to reverse it. It hopes to start the work of opening up this space to make way for radical alternative projects that can advance the agendas of anti-capitalism, feminism, trans activism, racial equality and other progressive projects.

Although not a heterosexual app itself, Grindr is a part of this process of digitised transformation of desire and having a videogame like *Hope: To See You Again* and Grindr under one roof is not merely fortuitous. Grindr is an early example of gamified dating, using location-based software now so common in games like Pokémon GO. Its comparable match count means matches with potential dates can be collected like coins or credits. As Evan Moffitt has recently written in an important reflection on Grindr culture, it has long since been common to attend hook-ups or sex events through Grindr where participants are glued to their phones, experiencing the event on the app itself while they are present live.[1] In that sense, Grindr was augmented reality before augmented reality, a gamified experience of sex and relationships that was ahead of its time. In videogame communities 'grinding' refers to the repeating of patterns of play to accrue points or experience, so perhaps there is a fortuitous clue even in the name. In 'Being Xtra in Grindr City', Gavin Brown writes that 'the app began to reshape and enhance our experience of the city's people and places'.[2] In a way Grindr was the start of augmented reality gaming.

This is a pattern of humans – and even their apparently deepest and most intimate desires – becoming predicted, influenced and 'gamed'. With our smartphones, smart condoms, sexrobots, dating apps, Fitbits, simulators and videogames, we are becoming increasingly robotic at the level of desire. When Allison de Fren studied the early online Usenet group alt.sex.fetish she noticed two patterns: those who desire robots and those who want to become robots.[3] Today, the two strands are stronger than ever, but there are also major differences. We are falling in love with robots more than ever before, and through patterns of gamification, prediction and engineering our very desires themselves are becoming roboticised. However, this no longer belongs to the realm of online subcultures or niche sexual communities. Now, both desiring and becoming robots are at the centre of social life. This book will claim that it is the normative position to be in love with a machine, not a quirky or interesting subversion of expected desires that belongs on the edgy fringes of social life. It will also argue that we are inevitably becoming more machine-like in our desires. We are robots, cyborgs, whose desires are changing with our technologies.

DESIREVOLUTION

In other words, this book argues that we are now in the middle of a digitally driven incarnation of what Jean-Francois Lyotard once called a 'desirevolution', a fundamental and political change in the way we desire as human subjects. Perhaps as always, new technologies – with their associated and inherited political biases – are organising and mapping the future. In today's 'anthroposcene' we can hardly survive a day without being reminded of the fact. What we are less attentive to, and what may be more unique to our own historical moment, is that the primary site on which this reorganisation of social life is taking place is libidinal. Our very desires are 'gamified' to suit particular economic and political agendas, changing the way we relate to everything from lovers and friends to food and politicians. Digital technologies are transforming the subject at the level of desire – re-mapping its libidinal economy – in ways never before imagined possible.

Developments in the digital industries are particularly geared towards infiltrating the spheres of love, relationships, friendships and sex (from sex robots and smart condoms to virtual dating, social media and hook-up applications) but even those less obviously connected to the realm of love (such as food and travel apps, videogames and self-driving cars, even election canvassing technologies) are implicated in a transformation of the subject at the level of desire. These transformations are deeply political and fundamentally economic.

This revolution in desire presents as many opportunities as it does concerns. There will be no 'humanist' argument here that we would be better off without or before our technologies. Desires have in some ways always been mechanised, set up and organised through social institutions and impacted and edited by technology. Whether through the church, the family, literature, film or the internet, desires have always been shaped and reconstructed by our institutions, and the situation today is no different.

What is different today is that an emergent form of what Nick Srnicek calls 'platform capitalism' has taken us into a situation of unprecedented inequality when it comes to the uses and development of technologies. Those in charge of our technologies and our data are an increasingly minute and increasingly powerful percentage of our population. If there was any truth in McKenzie Wark's idea that although we are invited to play god when we game, it is game designers who are 'gamer's gods' in that they design the world inhabited by the gamer – we are all gamers in a world of a very few gods – an increasing 1 per cent of those who set the terms and conditions of the world in which we play.[4] Because of this situation, much of what is found in this book might be considered negative or critical, but this is a battle that is not yet lost. In a world where technocapitalists, neoliberal tacticians, Silicon Valley machisimos and right-wing activists are directly engaged in organising the desires of the future for their own agendas (examples of which will be explored later in this book), we need to step up and seize the means of producing desire for ourselves.

Adapting quickly to our changing conditions is one way to start taking control of our new desires. Writing of the ways in which apps, sex tech and social media have transformed desire at the level of the body, Bogna Konior seems to extend Freud's idea that with technology man has become 'a god with artificial limbs', writing:

> Our bodies today are spread over a number of apps, each limb tended to by another wireless device, a piece of a body on the phone, a recording of a body on a website. … Some prefer technical erotics and await the arrival of sex robots. Machines spread our phantom bodies over the globe, opening it up to titillation, annihilation, de-subjectification, livestreaming us. Sexuality needs to adapt.[5]

Konior is also aware that this transformation of what it means to think, feel and desire is fundamentally connected to developing forms of capitalism. Imagining us all as robot lemmings controlled from above, she asks how long it will be 'until capital truly has the remote' and we become beings who respond instinctively to instructions issued out from a system that we created but which has run out of our control.[6] Tinder, for example, can be seen as an ally of a hyper-productive capitalism which allows us to organise sex into our neoliberal schedules and increase worker efficiency.

If my only desire, and thus my whole being [under neoliberalism], is to be an efficient employee, I have to move with the rhythm of capital accumulation. I have to become liquid myself if I am to mobilize for capital. I have to always be available and always ready to respond to the fluctuations of the market. Tinder, then, allows me to function as the perfect employee in a liquid market. I can choose to have sex at moments that do not hamper me [as a worker].[7]

This is just one of many ways in which the rhythms of everyday life are transformed to suit contemporary capital by a ream of new technologies each operating on a different part of our psyche to encourage us to become the subjects of the desired future. The French sociologist Henri Lefebvre introduced the idea of 'rhythmanalysis' to study the political changes to everyday life in the 1980s and a new kind of rhythmanalysis is needed in the age of platform capitalism and tech entrepreneurship to identify the changes in the patterns of our everyday life lived through apps, wearables and other extensions. Technologies of love, in particular, and our new patterns of friendships, relationships and sex, indicate this emergent everyday life as it develops.

In trying to understand how our desires, impulses and patterns of repetitious everyday life are changing, this book argues in particular for psychoanalysis as a way to make visible the new world of desire. Psychoanalysis – when it is at its best – is a way of thinking which connects politics with desire, making it a vital tool for the Left to deploy if it is to succeed in the war over the libidinal future. If this project is to be successful, however, psychoanalysis needs to be connected in new ways to both feminism and to Marxism. Often considered the remit of the elite and/or of men, psychoanalysis – as the study of desire – can be allied with feminist and Marxist theories of technology and of love to reveal the economic and political power structures that underpin even the most personal and apparently instinctive desires, drives and impulses that characterise social life today. Further, psychoanalysis has often been seen as focussed on individuals, whereas Ian Parker and David Pavon-Cuellar have shown that it can be an effective way of thinking *collectively* about our personal and political lives. A collective political approach to desire is what we need to deal with the technologies of today.[8]

In the 1970s, a group of Marxist and feminist-orientated therapists and mental health professionals put together a pamphlet called 'Red Therapy'. It argued that capitalism had begun to create new needs, new forms of 'consumer luxury, romantic love, sexual excitement' and other impulses. 'Parts of our lives that used to be controlled by religion (our sex lives, relationships, our personal and spiritual life) have now been invaded by the commodity ethic', they argued, and their project began from the idea that ideas and practices of psychoanalysis and psychotherapy could help us handle the situation.[9]

Things have come a long way in the last 30 years, but the idea of psychoanalysis as an ally of political activism is more important than ever. To have an influence of any kind on the technologies around us, operating through us and sharing our space with us, we need to understand the way in which they impact us at the level of psyche and of desire, and that has always been the remit of psychoanalysis. A psychoanalytic approach shows us how we are being reorganised and controlled by the new digital, corporate and political forces in society today.

There is a long history of connections between a critique of capitalism and psychoanalysis, in particular when it comes to questions of desire. In *Capitalism and Desire*, Todd McGowan describes the situation:

> Several anticapitalist theorists following in Freud's wake equated the destruction of capitalism with the complete elimination of sexual repression. They either worked to bring about sexual liberation with the belief that this would portend the end of capitalism, or they worked to combat capitalism with the belief that this would free repressed sexuality.[10]

Some key figures in this philosophical history have been Otto Gross and Wilhelm Reich, who believed that political and sexual revolution would go hand in hand. These theorists considered how revolution in the sexual arena could potentially lead to the elimination of repression in wider social life. This book is also interested in how the revolutions that are taking place at the level of sex, love and desire can be central to wider possibilities of social change. However, it does not work on the assumption that the end goal would be to 'free' or 'liberate' desire. Another theorist in this tradition, Herbert Marcuse, points out that 'the individual lives his repression "freely" as his own life: he desires what he is supposed to desire; his gratifications are profitable to him and others; he is reasonably and often exuberantly happy'.[11] In other words, capitalism appears to play a particular trick of making us experience our own organised, mediated and controlled desire precisely as if it is free, unrepressed and our own. Never has this been more so than in this digital age of technologies of desire. Since technologies edit and reorganise our desires to suit their own agendas, we can't trust our desire to simply work on our behalf.

McGowan's own work shows that it is not necessarily simple repressiveness that is the main characteristic of capitalism. For McGowan, 'the recurring fantasy within capitalism is that of attaining some degree of authentic belonging (in a romantic relationship, in a group of friends, in the nation, and so on)' but while capitalism 'spawns a type of fantasy, it constantly militates against the fantasy's realization' because if the subject were to reach the fulfilment promised by capitalism, it would stop needing to pursue the infinite pleasures and commodities of capitalism itself.[12] Capitalism doesn't just prevent us from getting or being what we want. It also creates the desire for what we want and for who we want to be, before mediating, limiting and controlling those desires. This wider capitalist logic is the backdrop for all the digital technologies that are discussed in this book. Our sexbots, simulations, dating apps, videogames and wearables all give us what we want, limit our fulfilment, control our desire and construct our impulses.

Although this book takes its inspiration in particular from the works of Sigmund Freud and Jacques Lacan, no familiarity with psychoanalysis is assumed here. The ideas of these historical psychoanalysts – along with the subsequent development of their work carried out by feminist and Marxist contributors to the psychoanalytic tradition – form the core ideas expressed in this book. Each chapter brings their collective approach to the politics of love to bear on a different aspect of new media and technological development. The first chapter considers the increasing role of data in the organisation of relationships. The second considers the smart city as the new space of desire. The third considers simulations of love from games to robots and the fourth discusses dating apps and the algorithms and interfaces which make decisions about who we interact with and how. The latter chapters finish with a playful 'pitch' which suggests how these technologies might be used differently with a progressive agenda in mind, were we to be able to wrest power from the few technocapitalists who currently dominate design and production. These are intended less as prescriptive answers to our problems and more as attempts to provoke discussion as to how we might repurpose our technologies and redesign our relationships to them in the future.

Each of the areas explored in what follows reveals different aspects of a new triangulation between politics, love and games. Throughout we will see that the Grindr saga is just a tiny little clue on the surface of a much deeper and more complicated set of connections between capitalism and desire in the gamifed world of love that we now inhabit.

1

Data Love

Politics is politics, but love always remains love.

– Jacques Lacan (1959)

A couple of years ago, with an almost imperceptible ripple, a story broke out in the French media reporting on the bizarre case of a young Danish proto-fascist who had become a kind of 'alt-Right' data miner. The misguided racist was using information harvested from one of IAC's biggest Match Group relationship websites, OkCupid, to support a fake-science argument in favour of an apparently new form of white supremacist eugenics. The story picked up little attention, and those who commented on it – in the progressive media at least – saw the attempt as a deliberate and gratuitous misuse of data to support a dangerous political agenda, which in some ways it was.[1]

Although part of a wider rise in dangerous right-wing 'race science', which the infamous Canadian psychologist Jordan Peterson can be seen as the soft edge of, there is something particularly important about data drawn from the realm of love. In this case, over 70,000 profiles had been scraped in the attempt to show a so-called 'cognitive dysgenics', ultimately aiming to produce a data-validated hierarchy of humans (through various methods including correlating intelligence with religion) with white people at the top. Other related arguments that have been made via dating data include proof that users are attracted to those of the same race, which is used to further support the logic of dating those just like you (upon which eugenics relies).

The importance of OkCupid as the data source in all this is not merely coincidental. Dating sites offer an unprecedented sample size of detailed information for miners, but in this case they are also seen as the means through which our species meets with a view to procreation: the place where dangerous eugenicist ideas could even be put into practice. It might be a far less extreme example, but we need only think of relationship platforms like Trump.dating or the Atlasophere (a dating site for readers of Ayn Rand, the philosopher and fiction writer idolized among the online Right and read by Donald Trump) to see how dating could play a role in eugenicist practices. It's perhaps not so bizzare that conservative journalist Toby Young, with worrying ideas of eugenics himself, should set up a Covid dating site during the coronavirus pandemic when parts of this book were written.

That scary prospect aside, what seemed to go strangely unnoticed in the OkCupid story was a more significant and even more frightening possibility. The question which threatened yet failed to arise was whether the data itself, and therefore the structure and algorithms of a major online dating platform (and perhaps of others, since Match Group owns dozens of sites which reach 190 countries in 42 languages) were in part responsible for the way the results had seemed to come out.

The racial biases of dating sites have been commented on before, but only in terms of how they *reflect* or *reveal* existing social biases. Instead, might it be possible that there is a more complex ideological code written into the history of the connection between love and the digital world – a connection which was already biased in favour of what might even be seen as right-wing and patriarchal identarian politics? OkCupid itself has commented on the racial biases of their data, blaming the fact that 'daters are no more open-minded than they used to be' and thereby circumventing the possibility that their own algorithms might be at least partially responsible for not only proliferating but re-writing sectarian trends in relationship building. In fact, the logic of matching users only with those just like themselves might be something shared between the dating algorithms and this extreme right-wing rhetoric.

Far from ratifying bizarre alt-Right pseudoscience, if this were shown to be the case, it would present the claims of the 'scientific' far-Right as just part of a wider and more concerning digital architecture that has become structural to relationships themselves, propped up by liberal assumptions about data's neutrality and innocence. In broader terms, the concern raised by the incident is that if relationships, desire and friendships are mutating and transforming in the age of AI, big data and social media, then there might be a problem with the politics they are mutating in the service of.

This question of what politics our technologies of love and desire are inheriting and proliferating, as is perhaps already evident, might lead us to have a fair bit of concern when viewed from a progressive or left-wing perspective, but we should also not look at this situation hopelessly. Instead, we might seek to make visible the political patterns in our emergent technologies with a view to changing them. Our very desires are being 'gamified' to suit particular corporate and political agendas, changing the way we relate to everything from lovers and friends to food and politicians. To succeed against these actors, we need a combination of economic and technological reform that would allow us to seize the means of producing desire.

THE LEFT AND RIGHT IN LOVE

Sometimes the content of a given technology in the sphere of love might appear be liberal-leaning, as with the long-running dating site and app Guardian Soulmates (closed in 2020), or even left-wing, as with the semi-humorous Red Yenta, the 2019 dating app for socialists. Far from being liberal or leftist equivalents of sites like Trump.dating, however, the logic of the interfaces and algorithms of these examples remains precisely the same as their counterparts. Wrapped in new or even radical packaging, for the most part these apps retain the structure of the traditional dating app. Here, the two components of the 'digital object' identified by media philosopher Yuk Hui are useful:

> There are two dominant forms of digitization: the first follows the system of mapping or mimesis (for example, the production of digital images, digital video, etc., which are visually and repetitively distributed throughout the physical world), whereas the second takes place by means of attaching tags to objects and coding them into the digital milieu (by means of this digital extension, the object then obtains an identity with a unique code and/or set of references). The second movement of objectification of data comes a bit later. I call the first process the *objectification of data* and the second process the *datafication of objects*.[2]

At the level of the 'objectification of data', the process by which we (as data) become represented as a digital object (in this case our dating profile), Guardian Soulmates functions very differently to Trump.dating. They objectify their users differently, at least superficially, with the digital objects showcased on one site almost the opposite to those showcased on the other. However, at the level of the 'datafication of objects', the way in which those objects come to relate to each other as pieces of data within sets and metasets, the process is more or less identical.

Guardian Soulmates and Trump.dating are not as different as they seem. This can also be seen at the level of ownership. Spark Networks owns both successful religious dating sites Christian Mingle (for the Christian community) and JDate (for the Jewish community), while also owning – and using similar algorithmic patterns to run – the site EliteSingles, a platform for wealthy daters. The difference between religious and wealth as a status appears only on the level of the digital object, its objectification, while it is ignored at the level of datafication. The politics of these sites, then, are not only found on their surfaces but in their shared methods of connecting (their algorithms) and framing (their interfaces) the experiences of desire which they construct.

These discussions pose a challenge to dominant assumptions about love itself, both in general discourse and in philosophical discussion. Love, we tend to believe, is apolitical. As well as in a great many other places, this conception is embodied perfectly by Bernardo Bertolucci's iconic 2003 movie *The Dreamers*, a romance set in the midst of the French 1968 riots. While Molotov cocktails are going off hourly on the streets, the characters, ensconced in their bourgeois Parisian apartment, are given vast distance from their context to experiment with sex and love. Rarely are love and politics discussed together, and a playful data-pull (in the spirit of this book) from Google Scholar articles shows that academic work on love has halved in the past five years as work on politics has exponentially increased. The more pressing our political struggles become, the more love recedes into the background, or so it appears at least if judgement is based on the academy.

By contrast, basing the judgement on popular culture seems to produce the reverse result: dating apps, relationship simulators and relationship-oriented reality television has exponentially grown in recent years, along with intensifying preoccupation with politics. However, while things like Trump.dating attempt to connect love with politics, the majority of popular depictions of both love and politics keep the two apart (*Love Island* at 9pm on one channel, *Prime Minister's Question Time* on another). 2020 and 2021 have seen a rapid spate of hugely popular TV shows about relationships and love (dozens of international editions of *Married at First Sight, Too Hot To Handle, Celebs go Dating, First Dates*, etc.). In all of these politics is completely stripped away, as if like in Berlotucci's *The Dreamers*, we turn to love precisely to get away from political turmoil.

Alain Badiou's *In Praise of Love*, originally written in 2009, offers some philosophical support to the assumption of love's apolitical status. Making a case for a separation of the two, Badiou explicitly says, 'I don't think that you can mix up love and politics.' Given the above, we might suggest that love and politics, gradually partitioned since the 1960s, have taken a turn towards reconnection. Nevertheless, a distinction that Badiou makes between an 'enemy' and a 'rival' provides a useful insight to the situation. In a long discussion, he describes the fundamental distinction between love and politics by comparing the 'enemy' of politics with the 'rival' in the sphere of love.[3] Badiou writes:

> The enemy forms part of the essence of politics. Genuine politics identifies its real enemy. However, the rival [in love] remains absolutely external, he isn't part of the definition of love.[4]

Here, it seems, 2009 could not quite account for 2019, where the political enemy of the Democrat or liberal implied by Trump.dating is coded as a structural presence in the potential relationship between lovers. The rival has become the enemy: not external to the relationship but foundational of the possibility of love. For Badiou:

> The word 'communism' encompasses this idea that collectivity is capable of integrating all extrapolitical differences. People shouldn't be prevented from participating in a political process of a communist type simply because they are this or that, or were born here or come from elsewhere, or speak such and such a language, or were fashioned by such and such a culture, in the same way that identities in themselves aren't hurdles to the creation of love. Only political difference with the enemy is 'irreconcilable', as Marx said. And that has no equivalent in the process of love.[5]

Badiou's comment strangely anticipates a data-driven approach to love that dominates the field today, and connects this with a critique of contemporary identity politics. Data here is on the side of identity. For Badiou, the politics of the Right might be irrevocably connected to identity politics, but the possibility of a communist political agenda would involve solidarity across identity categories. What we can add to Badiou

is that the implication of Trump.dating and the wider digital reconnection of politics with love is that the 'extrapolitical' is the only thing
excluded by love (the only potential match ruled out by the first part of
the site's logic is the political enemy). Nevertheless, this political enemy
is the very pre-condition for the possibility of desire between two Trump
supporting site users, making the logic a kind of triangular (and possi-
bly homosocial) bond (to borrow the terms of Luce Irigiray) in which
the third part of the triangle that makes desire possible between the two
lovers is the imaginary political enemy.

HOT RICH MEN

In the same year that Badiou wrote his book, Eva Illouz published her
significant contribution to the debate in *Why Love Hurts*. Illouz, one of
many feminist writers to emerge out of – and often in tension with – psychoanalytic theories of love and relationships makes Badiou's argument
seem as much symptomatic of patriarchal power structures in the discourses of love as it is critical of them. Illouz recalls a point highlighted
by feminists since at least de Beauvoir that there is a deep link between
economic (and thereby political) power and sexual power and ignoring
the connection can lead to a failure to understand how patriarchy and
capitalism work together to oppress.

> The most arresting claim made by feminists is that a struggle for
> power lies at the core of love and sexuality, and that men have had
> and continue to have the upper hand in that struggle because there is
> a convergence between economic and sexual power. Such sexual male
> power consists in the capacity to define the objects of love and to set up
> the rules that govern courtship and the expression of romantic senti-
> ments. Ultimately, male power resides in the fact that gender identities
> and hierarchy are played out and reproduced in the expression and
> experience of romantic sentiments, and that, conversely, sentiments
> sustain broader economic and political power differentials.[6]

This argument was earlier made by Shulmaith Firestone, who wrote in
her 1970 *The Dialectic of Sex* that 'a book on radical feminism that did
not deal with love would be a political failure' because love 'is the pivot
of women's oppression'. For Firestone, the 'omission of love from culture
itself' (still prominent today, as the trend of separating love from politics

discussed above shows) is a vital clue to its political importance. Arguing that examining the realm of love is tantamount to 'threat[ening] the very structure of culture',[7] Firestone counters Denis de Rougement's famous argument (made again with a twist in 2015 by Srecko Horvat) that there is a difference between selfish and narcissistic love and a more romantic equal kind of love. Firestone argues rather that:

> [Love] becomes complicated, corrupted, or obstructed by *an unequal balance of power.* ... The destructive effects of love occur only in a context of inequality. But because inequality has remained a constant – however its *degree* may have varied – the corruption 'romantic' love became characteristic of love between the sexes.[8]

In this formulation, there are not different *forms* of love nor necessarily always a distinction between love and desire, love and lust, love and infatuation, or any of the other divisions made both in general discourses around love and in theoretical discussions. Instead, love is operated on by power structures and transformed ('corrupted') by them. Here it's useful to think in the terms of *longue durée* (the long term) and *histoire événementielle* (evental history/the short-term changes driven primarily by events), terms set out by the Annales school of historians. For Firestone, the 'degree' of patriarchal power operating on the realm of love might have changed in the short term, but in *longue durée* terms has remained constant across large temporary and spatial distances. In many ways technological innovations of recent years belong in the realm of *histoire événementielle:* while they affect and change the ways in which we love, they also may leave intact longer and more entrenched power structures.

Building on this suggestion, Illouz's own argument counters the approach that love is always a reflection of power on the basis that such theoretical positions rely on the idea that power comes first and love follows, making love a kind of embodiment of patriarchal and capitalist power structures. While it might often be just that, for Illouz, love is 'no less primary than power' and so:

> reducing women's love (and desire to love) to patriarchy, feminist theory often fails to understand the reasons why love holds such a powerful sway on modern women *as well as on* men and fails to grasp

the egalitarian strain contained in the ideology of love, and its capacity to subvert from within patriarchy.[9]

Combining Marxism with feminist theory, Illouz sets out to explore love as a space for this kind of subversion from the perspective of anti-patriarchy. In many ways this is the battle in which this book takes up arms, since to fight against the control and reformation of love carried out by platform capitalism would also be to fight against its powerful inherited patriarchies. The question is whether these recent technological changes offer sufficient opportunity for reformation to affect the *longue durée* of love so that it could become a space of something in the service of solidary politics, gender equality and sexual diversity rather than an arm of state-approved heteronormative patriarchies. In many ways this has been the task of the Xenofeminism movement, which argues for an 'explicit, organized effort to repurpose technologies for progressive gender political ends', and seeks to 'strategically deploy existing technologies to re-engineer the world'.[10] Perhaps a new technology of love could even offer us a way out of a long history of oppression and control.

Gesturing towards how love and politics might help each other, the more useful insight that Badiou makes in his book is the conviction that politics *ought* to share something with love insofar as 'identities in themselves aren't hurdles to the creation of love' and ought not to be to the creation of political solidarity. In other words, if love can step across identity categories and form new connections, so too could politics. In *Mistaken Identity*, Asad Haider, speaking of Rachel Dolezal, the black rights activists controversially 'outed' as being born of two white parents, writes that

> Passing, in this sense, is a universal condition. We are all Rachel Dolezal; the infinite regress of 'checking your privilege' will eventually unmask everyone as inauthentic. No wonder, then, that we are so deeply disturbed by passing – it reveals too much to us about identity; it is the dirty secret of the equation of identity with politics.[11]

The existence of Trump.dating, or the so-conceived leftist equivalent Red Yenta, reveals to us this dirty secret that identity is equated with politics, not just in these cases but across the board. Those sites seek to achieve precisely the opposite of Badiou's hopeful suggestion: tying love into identity categories via filter bubbles and closing down the possibility of solidary across class and political identity categories.

Haider's point that we are all 'passing' could not be truer in the sphere of love, where 'passing' is the logic of the entire game: we might pass as a 39-year-old when we are in our mid-forties or we might exaggerate our career successes, or simply pass as a desirable object in a more general sense. Doing so is not to be subversive but to use dating applications in precisely the way we are invited to. We hear of those unmasked on a first date, or even after months or years of a relationship, with former or potential lovers 'outed' as other than what they originally claimed to be (known as catfishing). Users search for an identity and are primarily preoccupied with whether it is a 'genuine' one. The missing part of this puzzle is that this form of identity politics, so visible in the examples of online interaction discussed above, is not simply the result or symptom of cultural nationalisms and neoliberal multiculturalism that encourages us to think of ourselves in these terms. Instead, identity politics are irrevocably tied to data, and it is in the realm of love where this vital realisation is forced upon us. When we are turned into digital objects, when we are 'datafied', we have identity politics structurally imposed. To break this pattern, we'd need a different approach to data and how we use it.

SMART CONDOMS TO SEXBOTS

Ours is very much the 'data-driven' age. Arguments and claims made in the media and in the academy are backed up with harvested 'empirical' information drawn from data collection technologies that make dystopian cybernetic dreams seem like relics from the ancient past. Data sets control what we see on our internet searches, social media feeds and television screens, yet that process of selection remains deliberately obscured. Data as a methodology percolates through every area of the university, and new appointments (even in the so-called 'arts') reward scholars who apply data analytics software to understand everything from gender to poetry. Mobile apps manage everything from eating habits to menstrual cycles using data-formatted algorithms, while 'smart condoms' collect sexual movements into large aggregated sets which establish a new blueprint for the sexual future.

Data now determines what is normal. It organises us into patterns and serves as the dominant way in which we understand the world around us. The case of the smart condom says it all: it can collect sexual data for comparison with that of others in order to 'improve' sexual performance by directing it in line with the norm (or even towards bettering the norm at its own game and turning its users into some kind of data-ratified heteronormative stallions). But what does it mean for queer and non-normative sexual practices, for example, if new norms are created out of a data logic which by its structural nature rules non-conformist units anachronistic and out of time, determining them as 'not part of' or 'other to' its pleasure-logic?

This makes these data patterns part of a normalising process. Could such practices even be rendered *anachronistic* in the truest sense of the word (which contains the idea of *backward* in *time*), banished to the past or as belonging to the past by the normalising drive of data society which eradicates anachronisms from its data sets? Are we perhaps in danger of entering a new iteration of Foucault's 'great confinement' in which great swathes of data automate our very desires, bringing them into aggregating patterns which threaten to reduce 'madnesses' to digital silence?[12] Could such technologies be banishing misfit 'data points' from the 'data sets' that are used to construct the applications that go on to become the technological norms through which we govern and organise our lives?

If any of these questions prove even partially valid, it would take us into a future in which our 'artificial limbs' (to quote Freud again), the technologies that have become inseparable from consciousness itself, were constructed only on the basis of the so-called 'typical', with all its misogyny, racial prejudice and heteronormativity inherited by the technologies themselves. Such technologies are better seen as features of the cyborg subject rather than as simply mediators between fixed individuals. Rather than just thinking about means of connecting people, what is at issue here is the reprogramming of the desiring-subject, or of the loving-subject. It was Henri Lefebvre who wrote that 'computerized daily life risks assuming a form that certain ideologues find interesting and seductive'.[13] We need only think of the racist facial recognition software deployed from 2019, which was not designed to be such by its creators but which inherited bias from its data. Dating algorithms, of course, may similarly assume certain ideological traits from their creators, their initial data sets and their early feedback loops and then establish these as the norm.

Data claims to show us what is typical, but it also constructs the typical and makes it visible to us in a flash of understanding, where what is perceived appears to have been waiting patiently for our 'visualisation' to make plain. The very language of data, then, codes it as a mode of seeing and perceiving reality. We could playfully put this in the Kantian terms of 'transcendental schematism', which describes the process of translating an empty universal concept into something that relates 'factually' to our everyday lives. With all data, an abstract concept of what it might show pre-dates the visualisation, after which it appears to relate to our existence in a factual way. Using the idea in a way somewhat removed from the sphere of relationships and sexuality, Slavoj Žižek writes that 'it is at this level that ideological battles are won and lost – the moment we perceive [something] as "typical" … the perspective changes radically'. 'The universal acquires concrete existence when some particular content begins to function as its stand-in', he argues, and data can be seen as precisely such a stand-in for the search for the universal, ultimate truth, operating ideologically to make us believe.[14] Even if we doubt the data, it is already too late, because the typical has come into being, caught in the gap between doubting what we read and already having read it.

In the realm of love, this is a particularly vital point, and the smart condoms already mentioned are the tip of a very large iceberg. We don't need to see the code designed by IAC's technology to know that the logic is one of reflection or mirroring, connecting people on the basis of similarity and likeness (data has even 'proven' that people are attracted to those who look similar). The language of the 'match' is almost enough to make the point on its own. When Chris McKinlay hacked OkCupid to circumvent the logic of matching him only with those with shared interests, it was received as a charming tale of finding love outside of the algorithm, but was treated as a quirky anachronism rather than as something that called the whole logic of the site into question. Had McKinlay not been heterosexual, his act would have shown the data-driven approach for what it is. Such matching is only doing for relationships what Facebook does for friendships, connecting those considered likely to like each other (and also to *like* each other) on the basis of correspondence (in the original sense of the word). We act as if we would like to meet and converse with only those who are like us, because it has already become the typical way to relate to others.

What is important here is that this is not, as it might appear, the logic of culture nor of nature (whatever those terms are taken to mean) but of *data*. It is not sufficient to say that it is because we are a narcissistic society, for example, that our culture produces apps and websites that match us with those who are like us. It is also not because of an imaginary natural human drive that the data reflects racial separatism, the argument used in neo-fascist interpretations of such data. It is data whose laws, our society and our connections, friendships, and lovers now obey. Data is neither culture nor nature, and it does not reflect or reveal the truth of either. Instead, it is its own force, propelling us towards a continuation of the existing relationship between people and things, since it can only agree with the pattern and exclude the anomalous. Further, data can not only feed norms and continue them but produce new norms that have yet to be visualised as norms until the helpful data apparently assists us in the process. Data establishes and then extends or proliferates the typical, while also making it appear to have always-already been there.

At the same time as data-driven relationships become the norm, technology infiltrates sexuality in other comparable ways that at first appear subcultural, niche or non-normative. Often we give cultural or natural explanations for such bizarre occurrences, but they are also the product of data as well as of culture. Sex robots and virtual reality relationships, for instance (discussed in Chapters 3 and 4), reflect misogynistic and patriarchal traditions of domination and subservience (the cultural explanation), or else they gratify the infinite libido of the hungry sexual animal that is the virile male (the natural explanation).

In fact, both politically motivated interpretations or explanations of data-driven projects (left = cultural, right = natural) seem to neglect the role of data as a third force between culture and nature that operates additionally as an active agent in the process. The robots (and even dolls like those at LumiDolls in Moscow or PlayMate Dolls in Toronto) are constructed in relation to data patterns of what has been deemed to reflect the market. If Japanese men are our main audience and they statistically prefer younger Western women, that's what is built, or so the reasoning goes. Such is also the logic of Silicon Valley liberals who defend their work with the claim that their technology is neither political nor patriarchal but an innocent and apolitical result of data that indicates what people want, echoing OkCupid's circumvention of the role of their own algorithm discussed above. On the contrary, data does not simply show what is already there in nature or in culture.

Instead, such data-oriented developments do more than reflect what is already desired. For one thing, they code desire differently, presenting particular instances of desire as typical or universal and constructing desire itself (including that of those who deviate) in relation to those data-established norms. For another, they iron out inconsistencies or remove elements coded as unwanted or unpleasurable. The logic of data-driven dating sites, as well as sex robots like those designed by Realbotix and virtual relationships like those in *Summer Lesson* (see Chapter 3) (now available on the PlayStation Store in the US and UK) is not only to give the user what they want but also to exclude what is unwanted in the other. With Realbotix products, the user can 'customise' to remove any unwanted features of their sex robot and choose what 'personality traits' they want the avatar to have, not unlike an opening Plenty of Fish questionnaire (see Chapter 4). Thus, the process of giving the user what they desire is constructed as uniquely personalised when, in fact, it is deeply aggregated and based on excluding diversity in the sphere of relationships (from both each single relationship and from the wider data set) rather than proliferating it.

ARCADES OF DESIRE

In 1953, the ability of modern technology to transform desire was anticipated by the situationist Ivan Chtcheglov. In his seminal essay 'Formulary for a New Urbanism', Chtcheglov pointed out that new technologies idealise classically desirable situations (such as gazing at the stars, watching the rain, etc.) but exclude the 'unpleasant' things that are part of these fantasies. Our desires get cleaned up, so that only the nice facade shows. Iron and glass played this role in the architecture of the early twentieth century, making it possible – for instance – to gaze into a romantic vista of rainfall without getting soaking wet:

> The latest technological developments would make possible the individual's unbroken contact with cosmic reality while eliminating its disagreeable aspects. Stars and rain can be seen through glass ceilings. The mobile house turns with the sun.[15]

Because of this, Chtcheglov argues, we enter a kind of anesthetised malaise in which our desires have become more homogeneous, predictable, and automated by a process of cleaning or sanitisation. When desire is mediated through our screens it is cleaned up and edited to remove disagreeable aspects. Nevertheless, these experiences are charged with cosmic-like intensity and power.

Against those who have seen digital life as a world full of infinite or static desire, Dominic Pettman and Mark Fisher have described internet culture as characterised by boredom and disinterest. For Pettman, online social life is full of the 'homeopathic parcelling of tiny and banal moments' rather than desire or pleasure experienced as 'ecstasy or bliss'.[16] Chtcheglov's idea of a technological malaise or even automation which nevertheless has an extreme cosmic desire at its centre might be relevant in these discussions too. Does the catching of a Pikachu or the flick of swiping right on Tinder belong in the category of a bored disinterested grasping at desire or in the category of pure ecstatic bliss? Or is it strangely both?

Writing of how technology invokes a mixture of pleasure and loss, Yuk Hui introduces the concept of 'technological ecstasy', which he describes as 'a way of becoming that has no clear idea of its direction yet is characterized by acceleration and adventure'.[17] With technology, we experience a burst of possibility though the surface of the screen, even if it is not possible to determine what potentiality and possibility is contained within the moment. For Hui, this digital pleasure is not what Martin Heidegger calls 'temporal ecstasy', in which one grounds oneself in an authentic time but 'a hyper-ecstasy that celebrates speed while simultaneously being haunted by the anxiety of not being there'. Desire as we experience it digitally, we might say, is full of promise and infinite potential as well as of loss and absence.

In the year 2000 Clive Scott re-wrote Baudelaire's famous À une passante, the iconic poem depicting the rife desires of nineteenth-century Paris, describing the city of the new millennium as a new kind of 'city on heat'.[18] Today's smart city – the subject of the next chapter – might be better seen as such an automated malaise which is nevertheless driven by the extreme desire of its citizens on heat, frantically pursuing the desires of platform capitalism. Here, the digital spaces of desire might be something like the Parisian Arcades that Walter Benjamin described in his *Arcades Project*, a work that connects the psychogeography of Chtcheglov and Debord with the cities of Baudelaire.

Benjamin's ideas are particularly useful and interesting here because of the word *arcade*. Benjamin's *Arcades Project* focussed on nineteenth-century cities and the ways in which they were formed by new technologies, particularly the ironwork and glasswork that was used to produce immersive experiences in the heart of the city. In the nineteenth century we see emergence of these fantastical structures in Paris and in London (many of which are still visible today) which are the origin of the department stores and shopping malls. These huge glass structures are called arcades and they house within them all the products and promises of capitalism, from commodities for sale to experiences to indulge in. Benjamin discussed the experience of going into the arcade and becoming the perfect capitalist consumer because the subject is so affected by the technology of the building and its ability to immerse – immersion before technology was immersive in the contemporary sense. In this way, these spaces can be seen as a kind of nineteenth-century virtual reality.

Work by Oliver Grau and others has shown how architecture like religious churches and cathedrals can be seen as the origin of virtual reality. Stepping into these 'sacred' spaces, the subject enters a state of marvel under the ornate 360-degree ceilings and multicoloured glass windows, altars, isles and statues. Of course, this is the intention of a particular ideology – religious ideology in this case – to put its subjects into that state where they can be sold to, and the produce on sale can be anything from a commodity to a religion. Thinking of how the nineteenth century saw the forces of capitalism make use of immersive architecture, Benjamin considers the arcades as the ultimate space of immersion where people enter a semi-aware state, subjected to the power of the virtual environment and helpless to resist its charms.

It's not a coincidence, then, that videogames were also traditionally to be played in 'arcades'. Arcades are virtual and physical spaces which can be entered, which welcome us into their magical dream world where our desires can be fulfilled or where we are promised the fulfilment of every impulse.[19] Today, we might even say that the arcade has expanded to cover the entire city itself, or perhaps even beyond those limits. The arcade has no boundaries. Arcades are no longer temporary worlds of escapism or momentary hiatuses from daily life, but with our digital lives mediated through smartphones and our reality augmented wherever we look, ours is a global arcade of desire. Just like the churches of the religious world or the impressive glass and iron structures of the nineteenth century, this global arcade has its own political ideologies which are imposed on those immersed within it.

Whereas it was perhaps architecture that most significantly organised space at the time of Debord, Chtcheglov and the situationists, today we are organised at the level of desire by mobile technologies, location-based applications and data-driven algorithms and curation tools. Nevertheless, there is much we can learn about today from the situationists' attention to a revolution in desire that they saw as a key organising feature of modern capitalism. 'Dreams spring from reality and are realized by it', writes Chtcheglov, arguing that technological changes do not reflect existing desires (as OkCupid claims) but construct the future of desire. If our dreams become reality, we need to be even more suspicious of the data-driven patterns discussed here, which do not merely reflect what we want but take us into the future of desire. On the other hand, there is also a space for hope, and Chtcheglov argues that '[i]t has become essential to provoke a complete spiritual transformation by bringing to light forgotten desires and by creating entirely new ones. And by carrying out an intensive propaganda in favor of these desires.'

What is needed, then, is to take control over the process, asking what kinds of desires we might want to save from being banished into anachronism and which new forms of desire we might need to construct. The excuse of simply following the data provides a means for powerful actors to circumvent their involvement in such questions. Since data at the very least amplifies and extends normative trends (banishing anomalous points to anachronism) and at least potentially establishes new norms and patterns, data-driven projects are as political as they come. On the other hand, opting out of these dominant technologies offers nothing more than a technophobic nostalgia that will leave the progressive agenda out of the future.

In 2019, a New York-based cult announced plans to open another sexbot brothel in West Hollywood, only this time with a difference: it would be a queer and non-conformist attempt to subvert what have already become 'traditional' human-robot relations (those the data has encouraged). Some of their practices might strike most as bizarre, but at least this logic acknowledges that the robots are implicated in political patterns of desire and might even play a role in shaping such things, and that human-robot interactions (HRIs) are coded in political ways. Also in 2019, the Lora DiCarlo sex toy, the Osé personal massager, was banned from the Consumer Electronics Show, leading the all-female team behind the product to express concern for a future of keeping women and their influence on technology itself out of the sex tech industry. These patriarchal decisions impinge on the politics of the industry while conceiving themselves as apolitical.

In the future of love technologies, for couples and groups of all sexualities, data-driven approaches that obscure the powerful politics of their logic should be replaced by systems that admit that their algorithms are implicated in politics – sexual and otherwise. Sheila Jasanoff points out that

> many non-fictional accounts of how technology develops [as well as science fiction accounts] still treat the material apart from the social, as if the design of tools and machines, cars and computers, pharmaceutical drugs and nuclear weapons were not in constant interplay with the social arrangements that inspire and sustain their production.[20]

Recent science fiction, from Spike Jonze's *Her* (2013) and Alex Garland's *Ex Machina* (2014) to Hollywood and Netflix depictions in *Bliss* (2021) and *Altered Carbon* (2018–20), seems to have broken the trend and begun to recognise that the digital future is reconstructing what it means to love and desire along political lines. Factual accounts – from papers written by tech developers to media reports on technology – still overwhelmingly speak about their technologies as if they have nothing to do with the political whatsoever. Reversing this conceptual trend is the first step to repurposing technology for a progressive agenda.

In *The Radicality of Love*, Srećko Horvat questions that quixotic tendency to partition love from politics that has obscured these developments from most eyes. He argues that in recent years the Right has managed to manipulate the public's emotions much better than the Left, including love, and that – in a twist on Badiou's position – the Left needs to be more actively engaged with these libidinal transformations rather than backing a politics which risks ignoring the libidinal. For Horvat, new technologies in the sphere of love encourage a new form of narcissism connected to identity politics that rejects difference and otherness, ultimately favouring right-wing discourse.[21] Even if this bias is incidental, and not the deliberate intent of technology companies, Horvat argues that the Left will need a similarly new kind of political engagement with love to combat these trends. Projects on the Left in the last few years, such as the campaign for marriage equality in Australia and the backlash against the new porn laws in the UK, recognise that the political and legal world impinge on the personal world of relationships, but not that relationships are always-already political. But these movements must be accompanied by a more comprehensive understanding that love is always, and always has been, political. Such movements go some way to show that what we need is not a 'politics of love' but an attitude towards love that acknowledges its political nature.

SCREENS OF ENAMORATION

By 1977, Roland Barthes – perhaps the least in vogue of all philosophers of his generation at the time this book was published – had already acknowledged as much in *Fragments d'un discours amoureux* (*A Lover's Discourse*). The book is perhaps his most psychoanalytic text, which is significant in this context as it reflects his closest engagement with the problem of desire as political category. *A Lover's Discourse* is also – under the surface at least – an analysis of technology, even if technology must be understood as pre-digital in his own context. In a section entitled 'ravishment', Barthes discusses the moment when we 'fall' in love: a 'supposedly initial episode (though it may be reconstructed after the fact) during which the amorous subject is 'ravished' (captured and enchanted) by the image of the loved object'. Such moments, Barthes says, go by the popular term 'love at first sight', but demand the scholarly name 'enamoration' (an anachronistic term that Barthes adopts). For Barthes, enamoration is a 'hypnosis' that sets the subject on a path from which they cannot deviate. We suffer enamoration as an electrifying blow, inflicting in us a 'wound' that cannot heal until the object of desire is 'captured'. In other words, the object of desire dooms us to pursue it, even when doing so might be irrational and costly. When we are enamored, we 'need' the object in order to simulate a feeling of completion, though of course it is this mirage of fulfilment that constitutes enamoration's illusion.

This is Barthes at his closest to Freud, but it's also Barthes at his closest to Marx. In the course of this book I return to Barthes's work on the politics of desire several times with a view to show that his often critically neglected work in this area (in *A Lover's Discourse* but also across his career) can offer a vital counterpart to the combination of Marxism and psychoanalysis that is needed to think through the problems of politics and desire. In the above instance, we should take Barthes's remarks as referring not only – or perhaps not even primarily – to the lover but to 'the loved object', which is to say everything that works to consume our attention: our iPhones, our designer shoes, our sushi. Barthes refers to both the lover and the commodity, as well as the connections between them. The language of 'hypnosis' and of being 'fascinated by the image' puts commodification at the heart of falling in love. Although Barthes's criticism is directed towards an emerging world of commodities, he never engages in nostalgia for a love free from politics. Instead he describes a perennial process of consecration by which the object of desire is set on stage in a prepared scene for the 'subject' (us) to fall in love with:

> The first thing we love is a scene. For love at first sight requires the very sign of its suddenness (what makes me irresponsible, subject to fatality, swept away, ravished): and of all the arrangements of objects, it is the scene which seems to be seen best for the first time: a curtain parts ... I am initiated: the scene consecrates the object I am going to love.[22]

If the city – from the Parisian arcades immortalised by Walter Benjamin to the Champs-Élysées and Eiffel Tower of Barthes's own work – was once the stage on which people fell in love, then today's stage is the screen of the mobile phone. In this new technological theatre, we are all herded into position by complex algorithms and codes. Technologies have always structured the scenes with which and in which we fall in love, but new technologies have dramatically transformed these scenes, and, in the process, transformed love itself. Love, then, cannot exist outside of history, and will always play out against the political and technological set pieces of the day. While love 'at first sight' might seem sudden and instinctual, the scene has in fact been carefully set by the surrounding technologies, whether digital or otherwise. What Barthes asks is not that we reject this situation, but simply that we recognise it and pay attention to the forces that arrange and prepare the objects of our desire: the politics which construct the moment of falling in love.

In the next chapter we consider new forms of data-driven predictive technology and experimental augmented reality function as a kind of testing phase for Google's technologies of social organisation, organising, as they do, so many objects of desire in urban locations to maximise profit and to create a blueprint for the relationship between movement and desire in the smart city. Barthes would argue that we are dealing with much more than a profit scheme. Instead, technocapitalist corporations – Silicon Valley tech giants that acquire and organise more and more online space each month – are fundamentally changing the process of falling in love. As users of technology in the 'smart city', we fall in love with the Pikachu in Pokémon GO when he pops up on our screen and set off in desperate search of him, almost instantaneously. We fall equally hard for the image of a lover on Tinder or Grindr (which shares some program features with Pokémon GO), or even with the image of a meal as it appears on Seamless or Grubhub. As Barthes tells us, this kind of connection between technology, politics and love places the most power in the hands of those who set the scenes of our enamoration, who determine what and how we desire and how we respond to those desires.

While early proponents of the internet assumed it would be a democratising and heterogeneous space, it is crystal clear now that an increasingly small group of power holders control this new network of desire. IAC owns OkCupid, Match, Tinder, and over a hundred other companies, Google's umbrella company Alphabet acquired so many start-ups that they are able to influence the way we relate to everything from restaurants to Pokémon. Gaming corporation Beijing Kunlun Tech's surprise acquisition of Grindr discussed above – a formerly independent and at least arguably 'alternative' application – shows again that love is assimilated into wider patterns of tech development. Behind today's proscenium lie technocapitalists and a host of programmers whose products are increasingly subsumed by a small group of major conglomerates, centralising the space to program our screens and set our desires on their course.

AUTOMATION AND AUTOPIA

What these technocapitalists imagine can be described as an autopia. There are several reasons for the use of this term. For those advocates of the idea in Silicon Valley, it is a utopian dream of a perfect planetary world centred around the urban but reaching everything in the internet of things and automating everything into a seamless pattern of profitability without subversion. For others, autopia would be better seen as dystopian nightmare than utopian dream. Such individuals identify the concerns of such developing technologies of desire, perceiving how they might limit possibilities for protest and change. The limitation with this dystopian approach is that it tends to lament the loss of individual agency and fall into a more nostalgic longing for organic relationships away from the technologies of today.

There is another way of thinking through this problem, which is neither technophobia nor technophilia and neither dystopian nor utopian. Automation technologies are here to stay. Among the palpable examples are self-driving cars and trucks (which will take 3.5 million jobs in the US in the next few years), automated factory workers (already in implementation) and drone deliveries (trialling in Finland in spring of

2019), and these transformative moves towards automating could hardly be stopped, even if there was a will to do so. The vehicle industry is also at the centre of the imagined future, even without Tesla as the almost humorous exaggeration of this dream. Images of future cities focus overwhelmingly on cars and transport not so much because they are central to the future in any practical sense but because they serve as a visual image of it. The long history of the car as the object of desire – with all its stereotypical association with masculinity, wealth and power – is also not irrelevant. Our future is designed in tandem with the selling of an object and its image, and this process is inherently political.

On the subtler end of the spectrum, the automation of travel via Google Maps, as well as of running routes, shopping and music tastes by comparable predictive applications, the automation of conversation by 'smart compose' text and writing technologies and the gamification of love through dating technologies are likely to be irreversible. Neuralink, a company which specializes in software that will implant technology into the human brain, state that 'we will probably see a closer merger of biological intelligence and digital intelligence'.[23] While that might seem to some like a dream of the future, in fact many other technologies, which suggest the same merger of psychic and digital life, are already here.

Autopia, then, describes the present, at least in the realm of desire. Most importantly, what remains to play for is the politics of this automated world of impulses. It might be too late to live in a world without automation, were that even to be desired, but it is by no means decided who these technological changes to desire will serve. Autopia as a concept is an attempt to neutralise the judgements inherent in utopic and dystopic visions of the future and think of the digital future of desire as a battleground whose fate is as yet undecided. This book focusses on understanding the patterns that are emerging in the automation of desire and, as intimated earlier, it identifies in those patterns various strands of racist, misogynistic, patriarchal, anti-trans and anti-LGBTQ+ influence, as well as a broader tendency towards identity politics, sectarianism and the splitting up of potentially subversive and critical energies. However, it does not identify these trends hopelessly but in order to make visible and then plot towards changing them.

'The myth of love at first sight is so powerful', writes Barthes, 'that we are astonished if we hear of someone *deciding* to fall in love.' The association between desire and spontaneity, Barthes argues, is such a well-entrenched ideology that it seems impossible to escape. It may not be easy, or even possible, to take control of the process and decide when and how to love – when it comes to people or to objects. However, what Barthes's work calls for is an analysis of the scenes and screens in which enamoration occurs. Whether or not we *can* decide to fall in love, it has already been decided for us, the scene set by political and technological powers beyond our control. Horvat is correct that the political Right has been able to manipulate desires to serve their ends. We should respond in the way that Barthes laid out a long time ago: by rendering visible the politics of desire and recognising that technologies of enamoration now dictate our movements, organise the citizenry, and even influence elections.

Barthes's lesson is simply that love is always political, no matter how intoxicating and spontaneous it feels, and works to reveal the politics behind the scenes. With this in mind, after discussing the smart city as the broader scene of desire, this book considers in turn a range of specific digital developments in the sphere of love, from the chatbot, the sex robot, the virtual partner, the smart condom and the dating site, as well as less obvious developments in technologies of desire such as food and travel apps, videogames and self-driving cars, even election canvassing technologies. It considers each as a specific digital intervention into the realm of desire today. Rather than separating desire and love from politics and reason, it argues that we need more 'deciding to fall in love', a reconnection of the rational with the body.

For a historical example of Barthes's thesis about the politics of falling in love, we can do no better than to read Goethe's *Sorrows of Young Werther*, a title Barthes wrote in the marginalia of his *Lover's Discourse* but a connection he did not explore further himself. That novel contains an iconic love at first sight moment that embodies the way in which love connects to the organisation of objects. Indeed, it was perhaps the most iconic scene from any novel of its day, immortalised in engravings and artworks from the period and later. Werther, as Geothe's narrator, describes the moment of desire in the first person:

> I walked across the court to a well-built house, and, ascending the flight of steps in front, opened the door, and saw before me the most charming spectacle I had ever witnessed. Six children, from eleven to two years old, were running about the hall, and surrounding a lady of middle height, with a lovely figure, dressed in a robe of simple white, trimmed with pink ribbons. She was holding a rye loaf in her hand,

and was cutting slices for the little ones all around, in proportion to their age and appetite. She performed her task in a graceful and affectionate manner; each claimant awaiting his turn with outstretched hands, and boisterously shouting his thanks. Some of them ran away at once, to enjoy their evening meal; whilst others, of a gentler disposition, retired to the courtyard to see the strangers, and to survey the carriage in which their Charlotte was to drive away.[24]

The focus here is not on Lotte herself, of whom we learn only that she is 'a lady of middle height, with a lovely figure, dressed in a robe of simple white'. Instead, for Werther what is important is what Barthes would call 'the arrangements of objects': Lotte's relation to the children, the rye loaf, and the knife, all appear as scene-setting props which make desire possible. Lotte emerges from amidst these objects and Werther is 'initiated' as 'the scene' (described by Werther as the 'most charming spectacle') 'consecrates the object [he is] going to love.' It is that scene, that arrangement of objects, which makes desire – even love – possible. The technologies of our space, place and time set the scene for love to appear – make the emergence of desire possible. We don't fall in love with an object in isolation but with how it appears in a curate scene determined by a variety of technologies. The Tinder profile card could hardly be a more perfect example from today.

Including literature in a discussion of digital technologies of love might seem anachronistic. On the one hand, digital technologies and their ability to transform the world of relationships take us away from 'traditional' models of love and desire which are often thought of as being embodied by literature and film. But as Niklas Luhmann points out in *Love as Passion: The Codification of Love*, one of the most important historical contributions to debates around the semantics of love, when love is experienced or represented it does not appear alone in its context but as if in relation to the entire history of love as we know and understand it as a culture:

> Each individual characterisation of love must be understood as referring back to all others. As this is true of *every* characterisation, and thus holds true for *all the others*, every theme occurs in all the others as *the other of the others*.[25]

When we mediate and represent love, we also refer back to a long history of love and its imagery, semantics and representations. This point is intuitively true and relevant here: every advert selling a product or service in the realm of relationships or of sex, for example, doesn't operate in isolation but refers back to the existing series of metaphors, clichés and assumptions about love and/or sex as if 'referring back to all others', adding its own contribution to an infinite ledger of representations of love which influence and affect how we process the image in the moment of seeing it. Each appearance of love is, as it were, tacked onto the infinite preceding appearances. Like a blockchain ledger of collective memory, each experience and representation of love is added to the document, existing in relation to but also changing the history of things recorded there.

In this process, Luhmann suggests that literature in fact takes over in being the dominant mode in prescribing what love is, and suggests that literature even tells us how to perform love in accordance with its own logic, replacing social institutions like the family and religion which previously might have set the 'code' for how we love:

> The special sphere accorded to love relations makes it clear that here *the code is 'only a code' and that love is an emotion preformed, and indeed prescribed, in literature,* and no longer directed by social institutions such as the family or religion.[26]

This position shares much with the broadly psychoanalytic position taken towards love in this book. There may well be moments and experiences of love and desire which run spontaneously despite or even against the social and moral codes which organise and plan our experiences (an idea discussed later), but much of our experience of love and desire function as codified and prescribed experiences set for us to perform by a variety of institutions and structures. Today, it is literature's turn to be replaced as the dominant force in prescribing how we love. Today, it is not literature, the family or religion which dictate the codes of love but the internet, our smartphones and the network of online actors influencing the process. Love has always been codified, but the type of coding involved has changed.

LOVE vs LOVE

Writing about an important early incantation of the digital love object, the Tamagotchi, which operated as a kind of digital placeholder for a human or animal, Dominic Pettman makes the important observation that 'we are happy to say we love our dog, our car, our new shoes, our iPhone, our apartment, or even our country. But it is understood that this is *not the same kind* of love that we would have for our lover or spouse. That's the *real* kind of love.'[27] This distinction is often arbitrary, and serves the purpose of protecting or paring away one form of love, usually that endorsed by the speaker, from those others deemed due for criticism, saving an idealised form of love (often ironically accusing the other form of being just so) from those more obviously contingent on political and cultural trends. For Pettman, 'the pervasive romantic need to maintain a sacred space for "true love", untainted by commerce or calculation of any kind, ironically acts as a smoke-screen for the omnipresence of such forces.'[28]

Other recent interventions in discussions of love and technology, including Srecko Horvat's *The Radicality of Love* and Lauren Berlant's *Love/Desire*, have continued to separate love from love, albeit in more politically engaged ways than the standard cultural approach described by Pettman. For Horvat, while what he calls 'falling in love' can be seen as a commodified form of desire, *à la* Hollywood, there is nevertheless a form of solidarity or fidelity in love that can offer an alternative to those capitalist trends. Berlant partitions love from desire in psychoanalytic terms, and treats them separately. While love is an 'embracing dream in which desire is reciprocated', a dream which 'provides an image of an expanded self, the normative version of which is the two-as-one intimacy of the couple form',

> Desire describes a state of attachment to something or someone, and the cloud of possibility that is generated by the gap between an object's specificity and the needs and promises projected onto it. ... Desire visits you as an impact from the outside, and yet, inducing an encounter with your affects, makes you feel as though it comes from within you; this means that your objects are not objective, but things and scenes that you have converted into propping up your world, and so what seems objective and autonomous in them is partly what your desire has created and therefore is a mirage, a shaky anchor.[29]

In Berlant's terms, this book is perhaps primarily about desire rather than about love, since it considers all manner of objects, from humans themselves to food and other commodities (those that are mediated though technology) to digital objects of desire themselves from sex robots to Pokémon that seem to draw the subject into those often temporary 'clouds of possibility' that Berlant describes. Berlant's attention to the 'scene' of desire is also resonant with the approach here, though while she speaks of the 'things and scenes that *you* have converted into propping up your world' and 'what *your* desire has created' (my emphases), the focus here is on connecting these scenes to the world of platform capitalism, where our algorithms and interfaces prop up our world and are not so much reflections of our own desires as they are affects that 'make you feel as though [they] come from within you'. These impulses are less inevitable clashes that occur on the affectual boundary of our internal and external worlds but mapped and planned impulses almost implanted by platform capitalism into its subjects.

There is an unlikely precedent to Pettman's argument in the work of Freud, who also argued against the tendency to partition one form of love from other apparently lesser forms. In 1922 Freud wrote:

> The nucleus of what we mean by love naturally consists (and this is what is commonly called love, and what the poets sing of) in sexual love with sexual union as its aim. But we do not separate from this – what in any case has a share in the name 'love' – on the one hand, self-love, and on the other, love for parents and children, friendship and love for humanity in general, and also devotion to concrete objects and to abstract ideas. ...
>
> We are of opinion, then, that language has carried out an entirely justifiable piece of unification in creating the word 'love' with its numerous uses, and that we cannot do better than take it as the basis of our scientific discussions and expositions as well.[30]

For Freud, the connection between these different ideas of love in everyday language is suggestive of important connections between things we otherwise seem so keen to separate as distinct. This book takes this suggestion on and shows that it has a particular relevance in today's world of desiring digitally. Of course, there are differences between the way we love our children, our lovers and our Pokémon, but there are also important connections between the ways in which our technologies mediate and codify these forms of desire that make the similarities even more important than the differences.

Today, the *scene* of enamoration – that arrangement of objects upon which desire is predicated and through which it is engendered – has been replaced by or has merged with what we might call the *screen* of enamoration. It is in exploring our social media networks, unconsciously repeating our Tinder swipes, scrolling our dating platforms and bashing our gamepads and controllers that desire is born. When we find a desirable work colleague on Shapr, or a potential lover on Tinder or a date on Bumble, or even when we respond to a suggested friend on Facebook or follow someone on Twitter, the process has little to do with the subject-object desire relation in isolation. Instead, it is the arrangement of objects seen through the interface of these platforms and organised by the algorithms underneath them which prepare the potential objects of our desire for their consecration and our initiation. Even browsing Deliveroo for a potential dinner, the user scrolls through outlets depicted by one image, with a user rating, a name and two or three keywords, each choice experienced only in relation to the other potential choices, tastebuds 'stimulated' in a carefully constructed way even when the language we use to describe the process attempts to tie the experience to the natural impulses of the senses. Aimlessly clicking on Facebook, the social media page of a potential connection is profiled carefully in the sidebar on the page of your colleague, with an advert for an object from Wish that your cookies took an interest in last week displayed directly beneath it. Next to those images are others, carefully arranged in relation to the chosen one that leaps from the screen and sets desire on its course. The whole thing has been arranged with the 'sign of its suddenness' only loosely hidden. Such is life in the desiring digital city.

In 1987 Jean Baudrillard wrote playfully in *The Ecstasy of Communication* that the 'scene and the mirror' were to be replaced by the 'screen and the network'.[31] With the above in mind, it has never been more important to understand both the interfaces (screens) and the algorithms (networks) which constitute these scenes of desire.

2

The Digital Libidinal City

Too late, but she knew, this canny Parisian,
That love at last sight puts the city on heat,
Her motion elastic, her furbelows Stygian.
My whole self convulsed as she passed callipygian.

Clive Scott, translation of Charles Baudelaire's
À Une Passante (2000)

Following Bathes's argument that desire does not spring spontaneously from within but appears only when the subject and its objects are arranged in such a way as to make desire possible, we need to ask several questions. What are the scenes in which we fall in love? And perhaps more importantly – since this would bring in the politics of things – *how* are things around us arranged in order to create an environment in which love or desire is made possible? Before focussing on the screens/scenes of enamoration that constitute the endless range and variety of moments of desire experienced in digital culture today, this chapter considers the broader scene of desire: the digital city. If each encounter with our screens – from the Tinder swipe to the targeted ad – is a tiny moment of enamoration, a mini-ravishment in which various ever-changing objects are consecrated as those of desire, the smart city is the macro scene for these endless cycles of micro seductions.

DESIREVOLUTION IN THE EAST AND WEST

Far from the scenes in which most of us live and love, on the outskirts of Hangzhou in Eastern China sits Alibaba's Cloud Town, a Silicon Valley-style working and leisure hub specialising in cutting-edge research in artificial intelligence and smart city developments. Alibaba is one of China's largest and most important tech companies focussed on forging the future of the city, along with China Mobile and Tencent. Having long claimed to be apolitical, Jack Ma, the billionaire co-founder and executive chairman of Alibaba, was revealed in 2019 to be a member of the ruling Communist Party of China (CCP). This revelation is just another in a long list of links between corporate and state apparatuses that stretch far beyond the borders of China. Nevertheless, a glimpse into the projects the company is working on in Cloud Town, considered in light of these revelations, make visible the connection between smart cities and a planned future of state and corporate control with desire at its heart. Here in this dystopian tech village, libidinal drives – the mapped, predicted and edited desires of the city's inhabitants – are at the centre of smart city development.

Technologies in development at Cloud Town range from AI pedestrian crossing lights that use facial recognition to identify the age of a road-crosser and give them a longer green light if they are old/slow enough, to robotic city helpers who can be automatically deployed in cases of emergency. Alibaba's 'City Brain' project sees Hangzhou as the prototype space to explore new forms of automation that aim to turn the city into a smooth functioning utopia. As is well documented, this imagined smart city is a project that relies on precise prediction of inhabitant's movements and tendencies. Among leading tech companies, there has been a paradigm shift in analytics infrastructure, with 'predictive network analytics' overtaking traditional models of planning and older data-driven models. What is usually ignored is the complex relationship between predictive technology and desire. While the narrative of predictive tech is that it works out what we want before we know that we want it, a brief psychoanalysis of these libidinal processes will show that these technologies function to change rather – than simply respond to – our desires.

One of the projects underway in Cloud Town offers some suggestive clues and stands out over others as taking dystopian dreams of an automated future a step further. These are the AI 'drone cars' being built by Alibaba in collaboration with the car manufacturer Rover. The cars can 'respond' to the needs of their driver and passenger, collect drone photographs and footage for social media and connect to the City Brain to ensure the user's own ideal experience of the smart city. The greatest feature of the car, explained the proud representative on my own visit to Cloud Town, is that its media panel, linked to the user's smartphone, reads patterns of movement, food choices and potentially (in the future) even photos and comments, and then crosses this with millions of data sets to make predictions about what the user might like to eat and how they might like to travel there (or have the food travel to them). In other words, the unique selling point of the vehicle is that it knows when you are hungry and what you might like to eat, before you do.

In short, what this shows us is that the new citizen (as imagined by Alibaba at least) outsources part of their decision-making processes, and maybe even part of their desire, to the tech companies running the software on their devices. Our very impulses, whims and fancies (here approached via dinner choices) are mapped and planned in advance. As the employee implied on my own visit, the triangulation between data, predictive technology and desire could be the single most important relationship taking us into the dystopian smart city future.

This is not a unique case but a dominant pattern in tech development. These developments in Hangzhou are not so far from the technology of Google Now that has made headlines since plans for its implementation in 2012. Yet here a previously neglected aspect becomes visible: Alibaba could use its complex algorithms to privilege food outlets that use Alipay rather than those that use WeChat pay or have no e-payment system, for example. Likewise, if Google answers your questions before you've asked them, it also moves you down paths you might never have otherwise taken. It's the early implementation phase of a predictive tech revolution in which desire is not only predicted but incrementally edited to suit particular agendas – in this case largely corporate ones, even if Ma's association with the CCP has the alarm bells ringing too. Scholarship has for several years covered in detail the biases of news curation, psychometric profiling in advertising and the algorithms of search engines, but it has not seen this concretely in terms of desire.

Further, this is not an issue of corporate interests in isolation. The possibility made crystal clear by the revelation that Ma is now (and perhaps has been for some time) a formal member of the CCP is that this process of editing a population's movements and desires might not only be in the interests of corporate parties but of the closely associated state. Likewise, Alibaba's rival Tencent is another major tech company with close connections to the Beijing government. Its 'heat map' feature tells users where crowds of people are forming, but taken to its extreme such a tool could be used by authorities to prevent street protests. This is by no means the concern only of authoritarian China – a line our own media outlets endlessly repeat. It is significant, for instance, that former Google CEO Eric Schmidt was appointed head of a Pentagon committee designed to integrate Silicon Valley into the intelligence services. Whether it's Chinese smart cities or Russian bots, the language of the West casts Eastern equivalents in dystopic terms of influence and control, while upholding at least the semblance of a myth of a free and liberated digital space on its own shores. The reality is that these new patterns of the libidinal city of the future are emerging just as powerfully in the West as they are in the East.

While in some ways comparable with the technologies being developed here, what the Alibaba example more specifically seems to reveal is that there are a number of subtler and less direct ways to organise the movements and even desires of citizens in the urban future that are implicated in the emerging patterns of the smart city. Beyond surveillance and monitoring, there is a more libidinal transformation of the subject in relation to its environment. As they are used in the latest predictive applications, our digital footprints, besides *revealing* our desires, are also responsible for the very *construction* of these desires.

This is something intriguingly predicted by Baudrillard, when he imagined how the self-driving car might function in the late 1980s. Baudrillard looks forward to a car that would

> inform you 'spontaneously' of its general state and yours (eventually refusing to function if you are not functioning well) ... the communication with the car becoming the fundamental stake, a perpetual test of the presence of the subject vis-à-vis his objects – an uninterrupted interface.[1]

For Baudrillard, this automating car that would eventually become so much a part of its user as to be the kind of technological 'artificial limb' that Freud theorised, only here the object becomes not so much symbiotic with the subject as a cyborg-like cybernetic feedback loop but a persistent pressure on the subject to behave in a particular way with regard to his or her objects. Just like at Alibaba, more important than the idea that the car would drive itself is the conception that it might insist that the subject be ready to act – and ready to desire – in response to the car's suggestions.

While the initial phase of this technology might merely incrementally edit the user's libidinal relationship with food to benefit one corporation over another, the endgame may be a more ingrained pattern that the user finds difficult to break, if indeed they ever want to. Dinner suggestions hardly appear to be dystopian nightmare, but they are part of establishing libidinal patterns that can be 'nudged' to suit particular actors (more later) and then embed the subject in a libidinally motivated cycle which it can be guilted into sticking to. Wearables like Apple watches and Fitbits (not to mention smart condoms) – and the Q-ser (quantified user) communities who use them to actively edit lifestyle patterns – might also be seen as a pioneering part of this trend. Such examples might serve productive purposes individually, but they also show the changes occurring at the level of subjectivity in the datafied city – patterns of the future that are by no means formed with only individual citizens and their needs and well-being in mind.

Regarding the Cambridge Analytica scandal, Lee Grieveson has discussed how the company was centrally designed to better *predict* and *control* political responses of a population and 'marshal influence and persuasion'. By using algorithms 'built to predict people's "personality" and maximally *affective* for the purposes of "behaviour change"', he draws attention to the important implication of Cambridge Analytica's original tagline – now removed with their website – 'data-driven behaviour change' because it shows that predictive technologies are not only about anticipating behaviour but about actively changing the way its targets will act.[2] His perhaps fortuitous use of the psychoanalytic term 'affective' also provides another clue: the key to these changes are libidinal ones.

Games and applications that make use of the Google Maps back-end system or similar mapping application programing interfaces, or APIs (including Uber, Grindr, Pokémon GO and hundreds of others), should be seen as particularly important technological developments of the last decade in this regard. They are specifically complicit in these new regulatory practices and work to construct the new 'geographical contours' of the city, regulating the paths we take and mapping the city in the service of both corporate interest and the prevention of uprisings. More importantly, and more unconsciously, they are part of what Jean-François Lyotard once called the 'desirevolution' – an evolution and revolution of desire, in which that what we want is itself now determined by the digital paths we tread.

In 1981, Guy Debord famously wrote of the 'psycho-geographical contours' of the city that govern the routes we take, even when we may feel we are wandering freely around the physical space.[3] At that time, it was Debord's topic – architecture – that was the dominant force in reorganising our routes through the city. Today, however, that role is increasingly taken up by the mobile phone. It is Uber that dictates the path of our taxis, Maps that selects the route of our walks and drives, and Pokémon

GO that (for a summer at least) determined where the next crowd would gather.

Other similar map-based APIs dictate our jogging routes (Map-MyRun), our recreational hikes (LiveTrekker) and our tourist activities (TripAdvisor Guides). Pokémon GO attracted some publicity because it accidentally gathered crowds in weird places, but this should only alert us to its potential ability to gather crowds in the *right* places (to serve corporate interest) or to prevent the gathering of crowds in the *wrong* ones (to prevent organised uprisings, for instance). This makes the technology a stage more advanced than WeChat's heatmap feature, which can only reveal where crowds are already gathering. Such applications should be seen as a testing phase in the project of Google and its affiliated corporations as they work out how best to regulate the movements of large populations via their phones. Pokémon GO players were the early cyborgs, complete with hiccups and malfunctions – a beta version of Google's future human. These future humans will go where instructed, and their instructions are issued libidinally.[4]

While the electronic Pokémon or the 'in-game rewards' offered by many applications may not (yet) have the physicality of a lover who can be accessed via Tinder, or a burger that can be located via JustEat, the burger and the lover certainly have the electronic objectivity of the Pokémon. We can therefore see a transformation in the objects of desire taking place by and through our devices, so that we are confronted not only with a change in how we get what we want, but with a change in what we want in the first place.[5] The establishment of the new contours of the city, with all its rhythms and flows, is at the centre of this libidinal transformation.

These 'digital' objects are, for the most part, not objects which exist *only* in digital form but objects which relate to material objects found in the outside world. As Yuk Hui writes, with digital objects we are dealing with 'a new form of industrial object that pervades every aspect of our lives in this time of ubiquitous media.'[6] Examples of such objects include online videos, images, text files, Facebook profiles and invitations, among which categories are the specific objects mentioned above, from lovers to Pokémon. Some of these objects bear a special relationship to something outside of their appearance on screen. They exist – in the city, as it were – in 'real' life, only indicated by their appearance digitally on screen.

In what way do these digital objects that appear on screen exist in the off-screen world? To think about this involves dealing with a history of object relations theory. In his historically influential article 'Thing Theory', Bill Brown attempts to understand how we relate to our objects. He points to the interesting arguments on the topic made by Cornelius Castoriadis back in 1975. Castoriadis counters the traditional psychoanalytic conception of desire as being defined by the lack of a desired object, arguing that in fact the object must be present to the desiring subject as desirable, showing that the subject's psyche has in fact already fashioned it.[7] In other words, his non-psychoanalytic perspective (more aligned with the tradition of deconstruction) holds that the beholder of the object has always-already constructed the object in their psyche, perhaps best imagined as a space for the object to fill. For us to feel desire for something, we must already in some way be prepared to feel the pull of this desire.

Among media theorists, Gilbert Simondon's idea of 'information' can help us to understand this situation. For Simondon, information coming into existence depends not so much on the emitter nor on the message, but on a particular state of the receiver, which Simondon qualifies as 'metastable' because it is charged with potentiality so as to make becoming-informed possible. Combining these perspectives, we can perhaps usefully say that the information-object, the on-screen depiction of food, lover or holiday destination, appears to the subject for whom becoming-enamoured has already been made possible by its own set of political events and arrangements. What this means in our terms is that the scene of desire may be vital, but the subject must also already be in a state of preparedness so that their desires can be put to work in this space.

Speaking of how a psychoanalytic approach to our relationships to objects might differ from one driven by deconstruction, Brown puts it nicely in a footnote when he writes that 'deconstruction teaches that the word is never as good as the referent, but psychoanalysis teaches that the actual object is never as good as the sign'.[8] To digitise this: deconstruction teaches that the digital object is never as good as the real-world one it represents, but psychoanalysis teaches that the actual object is never as good as the digital information which represents it.

At the same time, digital objects like those that have become ubiquitous only since Brown's article was written in 2001 seem to offer another way of thinking through this relationship between objects and their signs. In cases of today's information-objects, both statements seem true.

On the one hand, the image on screen appears to pale in comparison to the original, offering only a taster or a glimpse of the pleasure that the 'real' object might provide the subject once acquired. On the other hand, the 'real' object can never live up to the standards of the image that advertises it, like the flaccid Wendy's burger that causes Michael Douglas's rampage in Joel Schumacher's *Falling Down* (1995). In other words, the relationship between the two apparent parts of the object – its material existence and its representative digital information – still seems characterised by lack, inadequacy and disappointment, at least when the experience is thought of in terms of a comparison between the real and the digital experience, whichever way around it is viewed (the real can fail to live up to the digital or the digital to the real). It might seem that this gap may appear to decrease the *pleasure* yielded from the engage-ment with the object (since it involves disappointment in one way or another), the relationship (or gap) between material and digital object seems to simultaneously increase its *desirability*.

This connection of desire and disappointment might be thought of as a version of how Lacan conceptualises the relationship between love and hate. In what he calls 'hainamoration', perhaps recalling Barthes's discussions of enamoroation above, experiences of desire can be charac-terised by an ambivalence between love and hatred. Lacan connects the production of this affect to Christianity, but perhaps it is another reli-gion that makes the most of this process today: capitalism.[9] We might conclude then that the gap between real and digital objects is the perfect ally of contemporary consumer capitalism. By promising pleasure and incorporating it with in-built disappointment, we move swiftly from one object to the next in the line of desirable commodities.

It is in this sense that we are now living in cities of desire, or as the quotation from Clive Scott's poem given as an epigraph to this chapter claims, that the contemporary city is the 'city on heat'. Dystopian depic-tions of the future city so often invoke images of pleasure, from the drugs and whorehouses of *Altered Carbon* (2018–20) to the sexual and violent spaces of freedom offered in *Westworld* (2016–) to the pastoral family bliss of *Downsizing* (2017). Such projections imagine the future city as a space that will respond to desire and provide a world of pleasure for its inhabitants, even if it must do so against the challenges of environ-mental and economic collapse. In the dystopian future we are heading towards, desire is not something the state must respond to but its very fuel, a continuous supply of which is required for the system to continue.

If, as per Marx's most basic interpretation, capitalism is based on the idea that production will produce more production, so too is it dependent on the fact that desire can produce more and more desire to serve as its resource. This desire drives even the lightest of taps or the briefest of clicks. We must be kept desiring and disappointed – in infinite libidinal loops – to keep this ship afloat.

#HUAWEIWARS AND THE LIBIDINAL CLICK

It's hard to believe it isn't material from a comic sci-fi novel or an unmade episode of *Futurama*. Chinese tech giant Huawei sends secret agents to the T-Mobile factory in the US disguised as engineers. They take pictures on hidden cameras and steal a small part of a robot named Tappy before escaping with a view to re-build it across the Atlantic. Later caught, the employees are disowned as rogue agents. T-Mobile sues Huawei. The American president gets involved, accusing China of unfair trade and intellectual theft on his favourite social media platform. Simultaneously, China's biggest tech firm Tencent, responsible for creating a censorship firewall for their own government, accidentally spend $150 million on American free-speech platform Reddit. Free speech advocates on Reddit protest against the investment, asking for the money to be sent back to China. China offers to return Tappy's forearm but not the Reddit shares.

Far from parody, this describes the events of February 2019, and points to an important political realisation about the status of today's technology wars and their representation in the media: this is not primarily about censorship and security. If it were, it could hardly have descended into narratives of such comic proportions. What lies behind these smokescreens of talk about international meddling and surveillance is a battle over the platforms on and through which our free labour and monetised leisure time, interactions and relationships are harvested and turned to corporate profit in a new era of 'platform capitalism'. Commentators have likened the US-China tech situation to a new iron curtain or even a cold war, but this focus on differing ideologies is a distraction from what is really a corporate battle for control over profitable populations in which both sides want more or less the same thing.

Having formerly produced only mobile accessories, as of 2019 Huawei have a 15 per cent share of the smartphone market, which is marginally more than Apple. They have been a leader in the development of 5G, and are in partnership with a number of UK and US companies with regard to coming internet infrastructures. 5G means more clicks and more rapid connectivity between sites and devices, and Huawei's share of the 5G terrain will see them benefit financially from clicks in those countries it operates in, as well as giving them some potential power over those infrastructures themselves. In the US, this advance is viewed with severe suspicion, with Senator Tom Cotton going as far as to say that Huawei is 'effectively an intelligence-gathering arm of the Chinese Communist Party'. The connections between the Beijing government and Tencent and Alibaba discussed above are not misplaced concerns, though the articulation of these concerns in US media does obfuscate the obvious parallels there.

Perhaps unsurprisingly, China is more willing to admit that this battle is really about the ownership of platforms and digital space rather than an ideology war over censorship and free speech. Indeed, its own firewall that blocks both Facebook and Google is rooted at least as much in an economic protectionism as in wanting to censor dangerous or ideologically subversive content from its populace, a fact Western media is intent on ignoring. The reality for most Chinese internet users is that they can easily access Western sites if they want to (many regularly do) but that the inconvenience of doing so ensures that the bulk of their valuable clicks remains within platforms from whom Chinese corporations profit.

The prevailing neoliberal internet myth that still presides in the US prevents them from simply admitting that these are economic issues of space ownership and forces them to cast the tech wars as a defence of freedoms when they are in fact predicated on the absence of freedom in the first place. This 'fear' of the Chinese tech giant is less about *whether* corporations should exert power over the means of communication, exchange and labour and more about *which* ones do, a reality China seems more comfortable with.

The backdrop for all this is a new world of digital work and leisure that represents one of the most lucrative aspects of what Nick Srnicek calls 'platform capitalism'.[10] Free labour and unpaid work have experienced a meteoric rise since platforms became central to the social fabric, from obvious examples like unpaid content creation on YouTube to grossly undervalued work outsourced via Amazon Mechanical Turk to more subtly monetised social media impressions which led Laurel Ptak to call for Facebook users to be paid. Christian Fuchs is among those to have pointed out that all of these forms of labour, whether 'unpaid, precarious, crowdsourced, informal [or] casual are milieus of ongoing primitive accumulation that feature high levels of exploitation.'[11] Nevertheless, it's the last form – the minutely monetisable click – which is at the centre of the new form of capitalism that Huawei and T-Mobile, and the US and China, are fighting over. This click, as we shall see, should be conceptualised as a libidinal moment of subject-subject or subject-object desire.

PAY PER CLICK PLEASURE

The indirect harvesting of valued-per-click leisure time by corporations has led many technocapitalists to support projects like the Universal Basic Income (UBI), which would free up users' time which could then potentially be spent generating valuable data and content on their own platforms. The driving force of this trend is the Pay Per Click (PPC) advertising campaigns that have grown simultaneously with corporations like Google over the last 15 years, but now the value of the click is not based only on the likelihood of purchasing success, as older models of Google AdWords and other targeted ad campaigns functioned. Instead, the click is conceptualised as a data-point that connects two or more actors in the network. It is those moments of connection between subjects and objects that have potential value to data-driven companies from corporate advertisers to election meddlers like Cambridge Analytica and policy influencers like Palantir. This only works because the user can be libidinally motivated to conduct the 'free labour' constituted by the click.

The situation was prophetically predicted by one of the most historically influential Marxists still alive, Mario Tronti. His 1966 book *Workers and Capital* gave rise to the concept of 'neocapitalism', which anticipates the environment in which the digital worker operates today. For Tronti:

> At the highest level of capitalist development, the social relation is transformed into a *moment* of the relation of production.

In this environment, the data-point connecting two people, generated at the moment of every click between social media pages, connects the social relation itself to a relation of production in real time. Seeing this in his own future, Tronti worried that society itself would run by the logic of the factory. Each interaction between individuals would incorporate a surplus value turned to profit by the class owning the means of social production. If the factory workers could be made to relate to each other in a way that was productive for the factory owners, so too could the entirety of social life be modified and edited for the profit of the capitalists.

> The whole of society is turned into an *articulation* of production, that is, the whole of society lives as a function of the factory and the factory extends its exclusive domination to the whole of society.[12]

It is this new 'social factory' over which Huawei and T-Mobile are engaged in battle. For each share in the infrastructures being developed, in this specific case the 5G network, comes a greater portion of the profits of the factory of the future, which is now indistinguishable from society itself. When the factory extends itself across society, its relationship to desire and pleasure must change. While in the traditional factory pleasure is excluded (to be had after work and outside of the factory) it now becomes a central resource. People must want to produce and enjoy producing moments of profit-generating production.

It is this network of desire that is being fought over by nation states and corporations, since that space represents ownership of the global social factory. One of the most important books on the internet written in recent years but completely undervalued is *The Stack* by Benjamin Bratton. Bratton shows that thinking cartographically about a two-dimensional international politics and technology fails to consider the complex networks of our digital architecture. We always think of international politics in terms of maps – from board games to mainstream media representations of the world. He argues that we should do away with a two-dimensional idea of the political world and makes the case for six layers of 'terrain', rather than just one. One of the distinctions he makes is between 'earth' and 'cloud', which can be a useful distinction for thinking about how politics plays out in the old days of physical territory acquisition compared with how it plays out now in the wars over digital space.

CLOUD DESIRE

While traditional international relations take place on the layer of earth (talk of borders, occupation, etc.), these corporate struggles operate at the level of cloud, and carry with them a different set of state and corporate politics.[13] In short, it is a new kind of space over which corporate states are fighting. Perhaps we can say that in Bratton's terms, cloud politics are being hidden under the smokescreen of earth politics. Our media is keen for us to think about US-China relations, for instance, so that it can obscure the deals between corporations that are really at stake in this equation. This might go some way to explaining why the differences between Chinese and American internal affairs don't stand in the way of Tencent wanting a piece of the Reddit cloud. Framing the situation in terms of iron curtain and cold war metaphors is part of an attempt to hide cloud politics beneath earth politics.

The fact that some of those involved in this piece of international 'trickery' at least appear to be acting in 'good faith' is also an important part of the formula. Trump was consumed with the importance of border crossing and had a fear of Chinese surveillance and infiltration, while Jinping likewise appeared to see things in traditional cartographical terms, often speaking of China's expansion into other parts of the global terrain. Psychoanalytically speaking, when a subject is structured on the basis of exclusion or abjection, the term coined by Julia Kristeva to describe identity that comes into being via the repulsion of an Other, there is more excluded than the subject thinks. Both Trump and Jinping are a case in point for such a subjectivity, as is clear from the most cursory of glances at Trump wall building rhetoric or Jinping's preoccupation with purging the Uighur population in Xinjiang. To gloss Kristeva in this regard, we might say that the unconscious mind excludes more than the conscious mind, whose exclusions have not really been excluded at all. Psychoanalytically speaking, the language of earth and terrain – whether of the wall or the Uighur communities – excludes not only those 'others' repelled from that physical space (those in fact who are needed for the subject to sustain itself) but also the other terrains not described by the language of the subject. Trump and Jinping's genuine obsession with traditional earth politics allows the real battle – over cloud space – to be hidden.

Trump and Jinping are not only the architects of the confusion between earth and cloud – they are also the symptoms of it. The worst part of this realisation is that those who were suffering most at the hands of Trump's foreign policy and Jinping's eugenics might not be suffering if this confusion were to be ironed out in their respective psychologies and those of their surrounding discourse. International 'earth' tensions themselves seem to serve the purpose of obscuring 'cloud' ones, which is the terrain over which the future is being fought. Further, in terms of relevance to desire, one of the things being obscured here is that this is no longer an international battle over terrain and space but a war over the desire economy of the future.

The cloud battle can be seen as a battle over libidinal space, in which both corporations and nations compete over the future of desire in and because of its connections to data. In this regard Bratton is again helpful. In his discussion of platform wars, he notes that Facebook (like many other platforms in the battle) is based on *artificial currency*. Bratton notes that Facebook has spawned a ream of 'twitch and reward' videogame applications like FarmVille which feed off the central data sets of Facebook itself.[14] Further still, the twitch and reward model of gaming more is than an off-shoot of platforms like Facebook: it is the very logic of those platforms themselves. Twitch and reward gaming is the logic that drives the libidinal economy of sharing, liking and connecting via minutely monetisable libidinal clicks that are built into an artificial reward system. When we hear discussions of the 'gamification' of social life, the focus is usually on apps like Habitica, a 'habit and productivity app that treats your real life like a game' in order to change the user's behaviour patterns. Such technologies show wider technologies of pattern manipulation that Debord would have been interested in, but we might go further and say that in a more fundamental way social interaction is 'gamified' as a twitch and reward social factory where the surplus value is creamed off by platform capitalists battling over this space, even if we aren't using apps like these.

This idea of clicking as a game-like libidinal experience flies against another long-standing misconception regarding the relationship between desire and data. As we saw in Chapter 1, one misconception is that data is used to help the subject get what it wants, while its role in having any actual effect on the subject's desire is almost universally ignored or else actively dismissed. Another equally significant misconception is that data, with its calculating nature of quantification, monitoring and regulating, is considered almost to be the opposite of pleasure. Data – collected into infinite spreadsheets and visualised in interminable graphs – is just no fun, or so we often think. Where pleasures are romanticised in terms of the body or the spontaneous mind, with their unpredictably, unknowability and momentary nature, data is all about prediction, computing and the apparently soulless quest to remove fun and enjoyment from everyday life. On the contrary, it seems that today the data must be right before pleasure can even emerge. Spreadsheets are sexy, after all.

Italo Calvino lamented the possibility that the body, with its tastes and desires, might be left behind in the future of the subject's relationship to technology. He wrote of the 'amorous relationship' that 'erases the lines between our bodies and *sopa de frijoles, huachinango a la vera cruzana,* and *enchiladas*'.[15] While in such a moment the food and lover become one in a kind of orgy of physical consumption, not in itself an uncommon literary trope, in the same novel Calvino warned of a time 'when the olfactory alphabet, which made so many words in a precious lexicon, is forgotten', and in which 'perfumes will be left speechless, inarticulate, illegible'.[16] Indeed, food technologies like the mobile application Mangia (a kind of Tinder for meals) and the interfaces of delivery middlemen like Deliveroo and UberEats exclude the olfactory by necessity from the moment of choice, even if the food still smells of something when it arrives. While Calvino is guilty of a nostalgic association of pleasure with the body, the phrase 'olfactory alphabet' suggests – in a way that closely echoes the ideas of Lacanian psychoanalysis – that the sense of smell is a language of its own complete with a politics and set of power relations. In a certain way Calvino unconsciously predicates the argument of Alenka Zupancic in *What Is Sex?*, that pleasure is not connected to the body as much as it is to language, the pleasure of talking about the '*sopa de frijoles, huachinango a la vera cruzana,* and *enchiladas*' complete with the cultural capital of their authentic names seeming to eclipse the bodily pleasure of physically consuming them.[17] At the end of the day, you can't taste a book either.

Beyond even that, there is a language of the tastes and smells themselves which constitutes the politics of taste, and it is at least worth nothing that this libidinal economy, what Calvino called the olfactory language, is the most excluded in the latest developments in immersive technology. In virtual reality, it is only the connected senses of smell and taste that are not directly involved in the affectual experience of the user, where the senses of touch, looking, sound, balance and spatial movement are at the heart of the experience. Needless to say, sex robots do not smell (though there is work on olfactory virtual reality in process), but this appears to be much less of a factor than we might think it would be. While there is a long history that has associated sex and love with the body and its tastes and smells, this code is now being re-written.

A world of electronic objects with no olfactory language – no smell – blurs the distinction between lovers, meals and 'in-game' rewards, all of which have distinctly different relationships to taste and smell but can be more comparable when they appear visually on the screens of desire. The effect of this shift is to increase the power of technological corporations by giving them a new sort of control over the way we relate to our objects of desire. Technologies have developed to have more power over sight and hearing than they do over taste and smell, so our desires must also be shifted towards a focus on those things that can be easily manipulated by our technocapitalists. The fact that younger and more online generations feel more averse to bodily aspects of sex than older ones is no coincidence – as we will see later in discussions of the porn industry.

If the boundaries between the way we search, desire and acquire our burgers, lovers and Pikachus are dissolving, it is not so much the old point that everything has become a commodity, but that this kind of substitutional electronic objectivity and libidinal connections between the ways we relate to people and things endows corporate and state technologists with unprecedented power to distribute and redistribute the objects of the desire around the 'smart city'. Pokémon GO, perhaps along with 'games' of love and friendship such as Tinder, Grindr and Shapr (whose card game-like swiping recalls the play of competitive cards or casinos), can be considered uniquely important in this future city on account of their pervasive nature. While traditional play, as Johan Huizinga and Roger Callois famously define it, can be seen as defined by limits, these augmented reality (AR) games (see Chapter 4 for dating apps as AR) have endless potential and perhaps their end-goal is to become the 'infinite interface' that Baudrillard describes, ultimately resulting in a gamified city in which our desires are not displaced into videogames but fully integrated into the game of urban life orchestrated by platform corporations whose libidinal capitalism relies on the prediction and editing of desire.

In the language of Facebook's boardrooms, this type of project is termed 'processed listening', the technique through which surveillance practices are integrated into personalised content distribution. Elinor Carmi's work on 'rhythmedia', the way media companies (re)order different components in a way that orchestrates desired rhythms and removes undesired ones has identified the ethical problem here. Facebook's 'nudging ordering' process involves ordering posts so that rhythms come into line with 'desired rhythms'.[18] In this sense, as Carmi argues, we should see these new digital patterns in terms of what Raymond Williams called 'planned flow' or analyse them using what Henri Lefebvre called 'rhythm analysis'. Obviously, advertisement bidding can influence this rhythm, and flows and rhythms can be edited incrementally to suit particular corporate actors. As Rob Horning puts it:

> What has been called pre-emptive personality or personalisation is how you get a certain package or information about what you might want that you haven't explicitly asked for from a commercial service.[19]

Alibaba's futuristic predictive car is a comparable piece of technology and it can be seen as indicative of a wider pattern of rhythm management that is ever more implicit in our daily use of more ubiquitous technologies, rather than simply as a showpiece smart car that most of us will never use. Curation, nudging and the algorithmic organisation of people and things work to shape our reality to suit particular agendas via a wide range of everyday technologies.

The realisation that what are usually called 'predictive' technologies are in fact interested more in 'nudging' and mapping out the new rhythms of the city than in simply monitoring or surveilling also suggests the need for a new conception of digital time. Armen Arvanessian argues that as a result of digital data

> time itself – the direction of time – has changed. We no longer have a linear time, in the sense of the past being followed by the present and then the future. It's rather the other way around: the future happens before the present, time arrives from the future. If people have the impression that time is out of joint, or that time doesn't make sense anymore, or it isn't as it used to be, then the reason is, I think, that they have – or we all have – problems getting used to living in such a speculative time or within a speculative temporality.[20]

Data technologies do not simply predict the future by guessing what an individual or group might do or want to do in the future. It is rather that those futures already exist, completely realised, and they reach backwards into the present to guide it. The possible paths for our desires to travel are mapped ahead of time by algorithms in the hands of platform capitalists.

The situation can be considered in relation to Freud's concept of the 'preconscious', a kind of interim space between conscious and unconscious thought. In Freudian thinking, the preconscious describes thoughts and desires that are unknown to the individual at the particular moment in question, but are easily capable of becoming conscious. It is impossible for unconscious drives to enter the preconscious without transformation, so the preconscious is not so much the inarticulate morass of unconscious drives but already articulate possibilities that may emerge into consciousness. As such, our mobile phones and their predictive technologies may bring into consciousness desires and drives which might otherwise have remained in the preconscious. These impulses were part of the individual's history and potentiality (their data, as the phone understands it) but they may never have emerged into conscious desire to be acted on. In other words, as users we now outsource an important part of our decision-making process to a device that is designed to map our actions in the service of state and corporate interest.

It may be, as Arvanessian says, that time is functioning differently in the digital world, but only those behind these transformative data-driven technologies see how this new time works. We could say that a kind of trick is played. Time appears to the user as linear, as if their desires existed first and their potential fulfilment followed, making technology appear as something that merely helps the user move forward towards what it wants. In fact, these technologies produce and edit desires and retrospectively implant them in the subject. It is only because desire can be predicted and edited that this new speculative temporality comes into being. Here, once again, we see that the libidinal space is the central area in which the battles for the future are being waged. Controlling that space is tantamount to acquiring the power to direct the subjects of the future.

DESIRE ELECTIONS[21]

As has become more visible since the Cambridge Analytica scandal, these techniques of nudging and manipulation can be used for more than just corporate sales. What are sometimes called 'dark patterns', tricks used in websites and apps that make the user buy or sign up for things that they didn't mean to, are tactics of the corporate advertising trade, but they have also become normalised as political strategy over the last several years.

Many were shocked that right-wing causes such as those associated with the Brexit referendum and the election of Donald Trump could garner decisive support via social media. These campaigns mobilised memes and fake news bots, but also used highly sophisticated targeted advertising based on these kinds of data analytics to brainwash and manipulate voting populations – or so the argument goes.

While many moderate liberals responded to this with an increased suspicion of social media and new technologies, in light of what has been discussed here it is clear that the Left needs not to *resist* algorithms, but to create new algorithms of resistance in place of those being used by the corporate state and those on the Right. To do this it needs an appreciation of what I have called the 'libidinal' nature of politics online and of the relationship between new data-oriented technologies and desire that have been discussed. The surprise result of the 2017 General Election in the UK, coupled with the popularity of Bernie Sanders and Alexandra Ocasio-Cortez in the US showed that digital media can be just as congenial to left-wing causes (though AOC would drift away from these in the subsequent years) as it has been to right-wing or populist ones over the past few years, but in order to put this potential to use, the Left needs a better understanding of how such processes work.

This could perhaps start with the revision of a number of assumptions about how social media affects people in the way it does. The similarities between the Trump presidential run and the 'Leave' campaigns in the UK referendum were not coincidental. Leave.EU, the rowdier unofficial 'second' Eurosceptic campaign, headed by Nigel Farage, had its communications director, Andy Wigmore, meet with Trump's advisors in the summer of 2015. Together, they discussed political style, in particular how Leave.EU could imitate the 'gaming' of media cycles achieved by Trump's deliberately outrageous remarks. Both campaigns kept their messages in the news by presenting themselves as offensively as possible, and then 'doubling down' when apologies were demanded, ensuring they were repeated in the news and on social media over and over again.

Subsequently, both campaigns recruited Cambridge Analytica, a company which promised to revolutionise political campaigning. Political parties have been using expressions of political opinion by people on social media as a campaign tool since at least the Obama campaign of 2008. Cambridge Analytica claimed to be able to take this a stage further. Using a method called 'psychometrics', the company inferred from 'likes' on Facebook, not just political allegiances, but specific personality types and emotional states, and then algorithmically directed tailored political content to their newsfeeds. This enabled them to reach the voters they anticipated would be most susceptible to their clients' ideologies. As one account of the method puts it, 'psychological profiles [can] be created from your data, but your data can also be used the other way round to search for specific profiles: all anxious fathers, all angry introverts, for example – or maybe even all undecided Democrats?'[22]

It is not clear quite how much Cambridge Analytica helped either Trump or 'Leave', and the likelihood is that this has been overstated. Both campaigns have been unforthcoming in acknowledging the company's help, and it may also suit the company itself to appear as a shadowy operator, capable of delivering the impossible. What is clear, however, is that stories about Cambridge Analytica have fuelled what many people seem to want to believe about social media: that it is turning us into uncritical zombies, reducing everything to the lowest common denominator, and brainwashing 'the masses', who are – anyway – always only a few steps away from outright barbarism.

The Left is already using forms of digital media to gain support online, but we will only be successful in this regard if we are able to move away from the view of 'the masses' implied in liberal criticisms of the Trump and 'Leave' campaigns as prone to confusion, thinking of such voters that 'they know not what they do'. Instead, we must recognise that social media users are now active political participants, not just a mass to be manipulated, and focus on how to understand and engage these users more accurately.

Fear of how easily 'the masses' might be manipulated stems from a very old conservative mythology, one that has a particularly vibrant life in the alt-Right's resentment-fuelled dismissals of 'normies' and 'sheeple'. The Left and liberal centre is not immune to this thinking either: witness how swiftly some leftists dismiss consumers of Fox News or Britain's *Sun* newspaper as simply stupid, or how some liberals speak of any group expression of political allegiance to the left of them as if it were a cult.[23] In a useful corrective, the literary critic Raymond Williams once remarked that 'there are no masses, only ways of seeing people as masses'.[24] Progressive forces cannot win if they make the mistake of copying the Right's belief in brainwashed 'sheeple' when talking about those who disagree with them or if they simply mimic the kinds of 'nudging' at the heart of Facebook's curation algorithms. If the discussions of digital libidinal data given above have given any sense of the complexity of the political relationship between technology and desire, then such an approach is obviously completely inadequate as well as needlessly divisive.

This is not, however, quite the same as thinking that people are free agents, and are immune to being influenced online. Success on digital platforms comes not from unscrupulously exploiting 'sheeple' by lying to them, or by manipulating the emotional vulnerabilities they may reveal in their online behaviour. Instead, as we have already seen, political success online comes from realising that digital media thrive on what psychoanalysts call 'libido'. While modern media have always worked by trying to trigger and direct our desires, online media depend on them in innovative and material ways, establishing the new libidinal patterns of social life today.

Another way of framing this is that both older forms of data-based advertising and traditional political polling asked the voter the question 'what do you want?', without sufficiently attending to the complexity of the question. According to psychoanalytic theories, the question of what we want is always bound up in the construction of desires through politics, and culture, and technology. This new terrain, then, cannot be negotiated without psychoanalysis. As Will Davies pointed out in the context of the UK Brexit referendum, 'Leave's greatest advantage was that it didn't have to specify exactly what was being left.'[25] In short, people voted for a sliding signifier loosely tied to a variety of things they might have indirectly wanted, without any attention to the construction and make-up of those desires in the first place. In the context of changes to the world of work, James Smith and Mareile Pffanebecker make the point in counter to some of Srnicek and Williams's optimism about the post-work age of automation:

> The ideological, economic and political measures they present do not take account of the problem of desire. Our desires are never simply our own, and therefore work can never be driven simply by 'our own desires'.[26]

Desires, psychoanalysis teaches, belong not so much to the instinctual body of the human, as if emanating from within, but to the boundary between the subject and its external world. Desires come from outside, but the subject experiences them as their own internal psychic impulse.

In a chapter of his 1922 essay 'Group Psychology and the Analysis of the Ego' called 'Suggestion and Libido', a title which seems to speak directly to the libidinal economy of Cambridge Analytica data, Freud discussed the relationship between individual and group desires. Using the concept of 'libido' to throw light on group psychology, Freud argues that a particular form of pleasure is found, not in pursuing individual desires, but in giving up what the individual wants for what the group wants.[27] This is far from the simple utilitarian argument that we must give up what we want for a greater good. Rather, it acknowledges that a new type of pleasure and desire becomes possible in the act of desiring with a group. The contemporary Right makes use of this form of political pleasure often. Trump's campaign was full of such creations of a 'desire of the Other'. Chanting 'build the wall' or 'lock her up' was as much about the pleasure of joining in with a desire so passionately enjoyed by others, as it was about any personally held policy conviction. The alt-Right constitutes another specifically digital example, where pleasure is particularly found in' grouping towards a political outcome. Far-Right leaning doxxing communities like lolcow (on which liberal targets are seen as cows milked for 'lols') and Kiwi Farms operate libidinally not because each and every user individually shares a desire to target the particular SJW ('social justice warrior') or trans individual to whom each thread is dedicated to harassing and threatening but because individuals' libido can be displaced onto this collective action.

'Why ... do we invariably give way to this contagion when we are in a group?', writes Freud. Ultimately, 'groups are distinguished by their special suggestibility', he continues, a point that might help explain the various distinctions between doxing sites, chan boards and more main-stream forms of social media (more on these in Chapter 4). It might further make sense of Facebook's own directed ad service, where psychometric profiling is not merely about who likes what but about the levels of suggestibility that define different groups who can be potentially mobilised. Freud connects this suggestibility in groups to the concept of love, arguing that it is at the level of love and desire that group psychology can be explained. As such, love is a vital component of politics:

> We will try our fortune, then, with the supposition that love rela-tionships (or, to use a more neutral expression, emotional ties) also constitute the essence of the group mind.
> First, that a group is clearly held together by a power of some kind: and to what power could this feat be better ascribed than to Eros, who

holds together everything in the world? Secondly, that if an individual gives up his distinctiveness in a group and lets its other members influence him by suggestion, it gives one the impression that he does it because he feels the need of being in harmony with them rather than in opposition to them – so that perhaps after all he does it *'ihnen zu Liebe'* ('for their sake/for love of them').[28]

The argument made by Freud here is close to one often repeated in summaries and paraphrases of his philosophy – that desire, or Eros, can be found at the root of all impulses and drives. In reality, this is something Freud rarely says. For Freud, 'love relationships', which can also be called 'emotional ties', are not so much the root or origin of everything, but the force which 'holds everything together'. We might be able to follow this suggestive difference and turn the assumption that in psychoanalysis love and sex are the root of everything completely on its head. It is not that love/desire/sex is the root-cause of all other cultural, political and social phenomena, but that Eros contains within it all cultural, political and social forces. In other words, it is not that we can explain cultural phenomena by looking at love, as generations of post-Freudian theorists have attempted to do, but that to explain love (which 'holds everything together') we must understand first the 'everything' within it – the cultures, politics and social factors which all come together in what it means to love, desire or be tied emotionally to something.

Jacques Lacan disputes the idea that psychoanalysis traces everything back to sexuality in the context of dreams. In a playful account worth recounting in full, he writes:

On the aeroplane back from Milan yesterday evening, I came upon a really nice article in a thing called Atlas, which is handed out to Air France passengers[.] It informs us that there are high-flying psychologists in America who have been investigating dreams. Because people investigate dreams, don't they? They've been investigating, and they've found that, in the end, sexual dreams are very few and far between. These people dream about everything. They dream about sport. They dream about a whole load of josh. They dream about falling. But, in the end, there is not an overwhelming majority of sexual dreams. Since the widely held conception of psychoanalysis, so we are told in this text, is to believe that dreams are sexual, the general public, which comprises the psychoanalytic propagation – you are a general public as well – is naturally going to be peeved. The whole soufflé is going to collapse, just like that, to sink to the bottom of the dish. Among this supposed general public – because all this is supposition – it's true that there is a certain resonance according to which Freud is purported to have said that all dreams are sexual. Except that he never, ever, said that. He said that dreams were dreams of desire. He never said it was sexual desire.[29]

To say that dreams are 'dreams of desire' does not only mean that they can involve wish-fulfilment, as Freud had argued. It means that for Lacan, apropos of Freud, dreams involve simulated desire so that they do not so much reflect *what* we desire as they reflect *that* we want to desire. Dreams allow us to experience desire as if it is our own. Among other things this could be usefully applied to videogames, where the gamer does not necessarily (indeed not very often) want what the game offers but nevertheless derives a yield of pleasure from experiencing the possibility of that desire. The gamer may not want to shoot a foreign invader in the service of American foreign policy, but they can never-theless yield pleasure from the simulated desire to do so.[30] In the realm of virtual reality and mobile dating simulators, discussed in Chapter 3, this psychoanalytic reading again answers the question that cultural studies struggle with. They are clearly not wish-fulfilment in the tradi-tional sense (no user really sees the cartoon avatar as the ideal object of desire) but they do allow the user to step into the libidinal relations between subjects and objects; to enter a scene of desire as if they are active agent within it. These Eros-like scenes of desire, then, contain all of their surrounding and contingent politics within them. Love is the most political thing going – so understanding the affects and drives of love means understanding the political relationship between people and things in their cultures and contexts.

Likewise interested in the politics of enjoyment, Samo Tomsic argues in his psychoanalytic assessment of libidinal economy, *The Labour of Enjoyment*, that:

What matters is that the use of language causes in the subject pain or pleasure and that this discursive action stands at the very root of the constitution, reproduction and intensification of power-relations. Power and enjoyment are inseparable.[31]

In those doxing communities, or when the chant of 'build the wall' is shouted at a Trump rally, a speech act is used to reproduce and intensify a set of power relations by libidinally operating on those subjects involved. While Left-alternatives to these far-Right digital spaces (such as 8Chan's/leftypol/board and Bunkerchan) and a more libidinal speech act culture in 'offline' politics (such as the 'Oh, Jeremy Corbyn' mock football chant) have emerged, the Left has often been behind the Right when it comes to putting this desire to work, not least because the Right's cravings often seem unpalatably close to fascism, taking forms the Left has been reticent to ape. Nevertheless, a movement from individual to group desire need not remain the preserve of the Right: there is no reason that the Left cannot harness digital media platforms and use group desire to assert a Left agenda. As this book continues, it will circle around this issue of a leftist or progressive conception of love, desire or solidarity (to use a few provisional terms here) in a variety of different contexts, looking for precisely how the work of winning the libidinal technology battle that seems to be at the heart of the digital future might begin. Certainly, they will have to take account of love as part of a political scene of desire, as a libidinal moment which brings together the political discourses of the theatre in which it is staged.

The causes that stand to benefit most from digital media are not determined by the terms 'left' and 'right', nor by 'honest' and 'fake'. They are determined by how much they manage to solicit users' desires. Political success online is dependent on two interrelated ingredients: data and shareability. Campaigns need a stockpile of contact information, records of prospective supporters' online behaviour, and innovative ways of interpreting them. Then, their material needs to lend itself to being shared beyond initial users, validated by the fact that 'ordinary people' are sharing it.

But only certain kinds of politics are alluring enough to provoke the casual online behaviour that allows campaigns to know where to direct their arguments. Digital media is a media of desire, and will always reward the political campaign with the most claim on the libidinal: in particular, campaigns have most to offer in terms of the collective pleasures described above. This is why the attempts of Activate – the Tory youth group in the UK – to emulate the success of the Left's Momentum was doomed to fail, even before they collapsed after a couple of weeks of infighting and scandal: digital-libidinal politics is not some pre-existing model that can be applied to any position or cause.

In the UK's EU Referendum, 'Remain' and 'Leave' used digital media to infer voters' likely political allegiances from their behaviour on social media, and used this information to inform targeting of social media advertising, to create mailing lists, and even to direct on-the-ground contact with campaigners. A version of this strategy had been employed in Obama's re-election campaign in 2012 run by Jim Messina, who was recruited by the Conservatives for the 2015 General Election. In that election, activists reported that Messina's model was especially effective in surreptitiously recruiting voters who – the model predicted – could be convinced to vote Conservative, without rival parties in the constituency even being aware that targeting was going on.

In the relatively mainstream electoral contexts of Obama's re-election and the 2015 UK election, Messina's model worked well, but it came under some strain in the EU referendum. The most straightforward reason for this failure is the fact that the model was designed to target extremely specific groupings of undecided swing voters in specific decisive parts of the country – much fewer in a General Election than in a referendum. Messina complained that by the time the data that was needed to make adequate inferences about groups' motivations was finally sourced, the campaign's spending limit (a restriction on what the campaigns were permitted to spend) had already set in. This delay in finding adequate data meant that Remain was limited in what it could do with it when it was finally found.

More significantly, the Leave campaigns could draw on years of data trails, networks, and online interactions occurring in the orbit of the Eurosceptic party, UKIP. This was material people had long been sharing online. Allegiance to the EU, by contrast, had historically created no such 'libidinal institutions': nothing like the culture of Facebook groups, shareable memes and 'likeable' pages that are producing hordes of data. Digital media, in this case, was necessarily on the side of whichever politics was best suited to creating these stores, material that people took pleasure in spontaneously sharing. It was not a question of which side was most willing to lie and brainwash people, but of which side tapped into these libidinal behaviours the fastest and most effectively. It's no surprise in light of this that the Silicon Valley tech types of Facebook, Google and Alibaba have been the first to benefit in the emergent digital libidinal city.

Against expectations, however, these same 'libidinal' patterns served the Left spectacularly in 2017. After Leave's successful use of lurid ads about the risks of remaining in the EU, the Conservative Party, under Messina's guidance, spent millions on Facebook advertising focussed on Jeremy Corbyn's alleged terrorist sympathies and Labour's economic irresponsibility. The Labour Left campaigning group, Momentum, meanwhile, was criticised for its irreverent memes, and videos satirising Tories, and was accused of 'preaching to the converted' with short clips of Corbyn's rallies (the same criticism continued to be made after the election about Corbyn's appearance at Glastonbury). At the same time, Momentum created a 'My Nearest Marginal' app, allowing Corbyn supporters to bypass local parties and simply turn up to campaign in the places it would make the most difference. The Labour Party itself adopted another Momentum innovation – the Labour phone app, which allowed any party member to talk to undecided voters in their own time, and from their own homes.

Momentum's digital strategy was vindicated. While no pro-Conservative election material 'went viral' during the campaign, Momentum's did regularly, because it was funny, and created collective pleasure. Indeed, backing Corbyn online became its own pleasure, as it meant being welcomed into a culture of irreverent memes calling him 'the absolute boy'. The apps were also a 'libidinal' pleasure to use, because in contrast to the self-sacrificing and technocratically controlled atmosphere of traditional Labour campaigning, this was to be done on your own time, with your own friends, with technology that was second nature. All these examples of Left technology involve empowering users, rather than conceiving of the target user as passive and susceptible to manipulation.

The broader Left's realisation that appealing to 'the people' now means appealing digitally to people's desires – in all their complexity – can be witnessed in various examples of emerging Left technology. The New Inquiry's 'White Collar Crime Risk Zones' application, for example, uses industry-standard machine learning to predict where financial crimes will happen across the US, combatting the right-wing trend (and policing algorithms) which associate crime statistics with impoverished and non-white communities, instead showing a link between wealth and criminality. Similarly, The Southern Poverty Law Center has created a 'hate group tracking map' identifying fascist and racist organisations in the US and allowing users to follow their movements and organise resistance. In each case, the technology enables users to enjoy taking an active role in its potential uses.

Left-wing 'memeing' is another example of people putting themselves to work in the service of a political movement, extricating memes from their apparently apolitical history and seeing them as important political tools. As Matt Goerzen observes:

> Beyond the musings of think piece writers, memes are now taken with the utmost seriousness, by entities ranging from DARPA US military researchers and NATO agents to ISIS's ideological warriors – all of whom see the form as a contemporary weapon of war.[32]

The kinds of actors taking libidinal phenomena like meme seriously ought to set the alarm bells ringing. Likewise, companies like Palantir, who have close associations to US military figures, use data-driven solutions and many of the tactics discussed here to 'transform the way organizations use their data'. Their services are 'deployed at the most critical government, commercial, and non-profit institutions in the world'. In this context, it's more important than ever that the Left be involving itself with data-driven and libidinal futures. The question of an internet of desire – at the heart of this book – is also a question of the political future.

The many programmes created by the hacktivist Aaron Swartz form trailblazing parts of this evolving culture. Swartz commited suicide under immense pressure from the US courts for 'stealing' intellectual property when he attempted to make material gatekept by universities available to the public for free. Among other things, Swartz was responsible for campaigns for net neutrality, online petitioning and lobbying sites, as well as the creation of the Creative Commons, before his premature death under pressure from a US government determined to make an example of him. Recently, Evgeny Morozov and Nick Srnicek have made crucial appeals to the Left to become more involved in imagining alternative ownership models for online platforms like Google and Facebook that could wrest these essential tools of daily life from the self-interested and secretive bodies that currently control them. Corbyn's own shadow cabinet proposed a state-provided wifi service, which the British population repelled for fears of surveillance. Private companies are trusted more than the state or each other, and this situation has to change.

All these examples of Left technology involve empowering users as active participants and investigators, asking them to search, share and create online 'content', rather than conceiving of the target user as passive and susceptible to manipulation. Such technologies recognise users as libidinally motivated actors, not as brain-washable sheeple. If social media users were not 'sheeple' when they were enthused by Corbyn online, neither were they 'sheeple' when they were won over by Leave or Trump. A tech-savvy Left will win again by continuing to develop innovative digital media for campaigning: but it also needs to commit to a Left-'libidinal' understanding of what audiences are when they use digital media, what convinces them online and provokes them into action. While traditional politics has claimed to keep itself apart from pleasure and desire, if the Left is to succeed at this juncture, it must recognise not only that desire is influenced by politics, as psychoanalysts have long argued, but that politics must sometimes be influenced by desire. This section – something of a departure from the focus on romantic love – shows that understanding the love industry is also part of understanding a much wider shift in the libidinal economies of today, which affect everything from relationships to elections. The future of our cities, from our governments to our daily behaviour patterns, rests on understanding the relationship between technology and desire.

LOVE AT LAST SWIPE

While Barthes saw love as deeply political, and might be likely to agree with these interpretations of the scene of contemporary desire, Baudrillard takes a rather inverse position. He intriguingly sets up 'seduction' as that which might fight against or oppose the surveillance state. Writing on this strangely psychoanalytic aspect of his work, Isabel Millar writes:

> For Baudrillard, seduction is the last defense against the oncoming age of simulation, artifice, surveillance, computation, and ever more sophisticated methods of biological and molecular control. He asks 'how does one disguise oneself? How does one dissimulate oneself? How does one parry in disguise in silence in the game of signs, indifference in a strategy of appearance'? He affirms, it is not the desire of the subject anymore but the destiny of the object which we must be attentive to.[33]

The last aspect of our lives to fall prey to forces of simulation, artifice, surveillance, computation and control would be the process of seduction Baudrillard imagined in 1987. In the smart city of 2022, this seems to be almost the reverse of where we stand in the relationship between desire and the surveillance state. Today, seduction appears the crucial means through which the artifice of tech companies, corporations and governments can be carried out. For Baudrillard:

> In an amorous seduction, the other is the locus of your secret – the other unknowingly holds that which you will never have the chance to know. The other is not (as in love) the locus of your similarity, nor the ideal type of what you are, nor the hidden idea of what you lack. It is the locus of that which eludes you, and whereby you elude yourself and your own truth.[34]

In this strangely psychoanalytic moment amidst his critique of psychoanalysis, Baudrillard shifts the focus from the desiring subject to the object around which the subject is structured: something close to what Lacanian psychoanalysis terms the *objet petit a*. In the psychoanalysis of Lacan, the *objet a* is seen as both the elusive object of desire for which the subject endlessly searches and also as the object-cause of desire, that which sets desire on its course and makes desire happen, driving the subject to keep desiring and seeking its objects in search of this elusive ultimate object. In this way, this part of love is indeed the opposite of surveillance, computation and control because it is by definition elusive, unknowable and unmeasurable. While predictive technology and data-driven models of marketing rely on working out what the consumer wants, in a certain sense their ultimate desires remain unknowable.

Baudrillard here seems to critique directly – in a way just as anticipatory as his comments about the self-driving car – the quantifying institutions of online dating as only being able to offer 'false' substitutes for this more fundamental and knowable object at the centre of the subject's desire. Those formulations work precisely by the locus of similarity and the ideal type, or at the very least as offering the lacking piece of the subject's identity. Such approaches as online match-making services are doomed to fail, Baudrillard might argue today, at least in the ultimate attempt to exercise molecular control over love itself, because ultimately they can only provide the subject with a locus for their similarity (*à* la Trump.dating), or perhaps the idea of what they lack (as perhaps is what Tinder purports to offer in its role as a stop-gap for any moment of boredom). For Baudrillard, the ultimate form of seduction is found elsewhere, in the endless lure of a secret that the subject cannot by definition know. The point, then, is that the dating app does not actually want to lead the subject to fulfilment. Contrary to the ad campaign of Hinge ('Hinge is designed to be deleted'), the app benefits from the subject being in a loop of unfulfilled desire for substitutional objects of similarity.

However, Baudrillard is doing what many others have done before and since in partitioning love, of a certain special kind at least, from the political. Similarly, Lacan, who is invoked here, had said that 'politics is politics, but love always remains love'.[35] Indeed, the *object a* as a concept could even be seen as such an attempt to separate superficial desire for stand-in objects from another more fundamental desire at the heart of the subject. Yet, in the digital context of today, maintaining this separation seems to serve only the logic of corporate capital. Politics has succeeded in getting hold of – or has always had hold of – even the most apparently abstract and elusive seduction. In fact, this infinite cycle of replacement objects which stand in for the *object a* could be seen as the motor force of contemporary digital capitalism rather than that which escapes it. Lacan's argument, then, is not a description of the subject beyond politics, which it has been mistaken for, but an explanation for the workings of the political subject. At the same time, the popularisation of psychoanalysis as the theoretical support for the idea of the subject as infinitely and endlessly unfulfilled, the subject who must be infinitely seduced by stand-in objects in attempts to grasp at elusive pleasure has served to further justify that capitalist logic. In other words, a mis-deployment of psychoanalysis has forced it on to the side of capital. Desire operates in the way Lacan describes, but only because of capitalism and not inevitably so.

This chapter began with a quotation from Clive Scott's translation of Charles Baudelaire's *A Une Passante*, a poem which depicted the desires that for Baudelaire characterised urban life in the mid-nineteenth century. Scott's imaginative re-writing, which describes the subject convulsing with desire at the passing object, seems to look forward beyond his year of writing in 2000 to anticipate the desires of the smart city. 'Love at last sight puts the city on heat', writes Scott, framing the city as the space of endless moments of seduction, endless economic exchanges between subjects and their objects at almost every click which keep us swiping, buying and dating in the new patterns of the smart city of desire. It's telling that most major dating apps based on swiping, such as Tinder and Bumble, recently added a 'reverse' or 'undo' feature so that you can turn back to the previous card (often at a financial cost of a micro-trans-action). Desire for the object just gone is sometimes the strongest of all, and that has been monetised accordingly.

The Left today must work against issues of surveillance, artifice and control in the realm of love, as Baudrillard argued. However, it must do so without falling into the trap of separating love from politics in an attempt to protect some elusive remaining form of desire or love from infiltration by those forces. Instead, we should follow the suggestion from Freud – the reverse of that which is usually attributed to him. Desire, love and sex are not the fundamental forces that drive subjectivity over and above all other factors from the political and economic to the social and cultural. Instead, those apparently 'deepest' impulses and instincts (which Freud preferred to call drives) are the very site on which politics and the subject come together. Freud wrote that Eros 'holds everything together' and this should be taken as meaning that it is the very fact that desire includes the political, cultural and economic that makes it such a powerful site of power. What comes from Barthes and from psychoanalysis, then, is that love is and must necessarily be political.

Seeing an impasse in the political Left and conceptualising this in relation to politics and desire, Cindy Zeiher describes the 'rise of an ineffective postmodern-liberal Left which no longer requires action in order to appear political'. For Zeiher, 'in spite of its posturing and acting out, today's Left lacks the presence of a motivation which centres on the act as the primary location for entry into the Other'.[36] The contemporary Left is insistent on defining itself as political, but does not need to act in order to appear so. Against this, she argues for an act that is 'a point of departure for the subject in that the act combines desire, demand and sacrifice in an explication of political subjectivity'. To be described as political, then, the subject must act in ways which implicates desire and simultaneously involves both demand and sacrifice.[37] Many of the libidinal moments discussed in this chapter have been new forms of digital act that might fall into this category of defining the political at the level of desire.

PITCH – RED WEARABLES:
THE POLITICISING SMARTWATCH FOR WORKERS

There is no reason why the new digital patterns of our lives should not work effectively and powerfully for the Left or for progressive forces and against the corporate, Right-populist and occasionally liberal interests who they seem to have served so much in recent years. As mentioned, when Jeremy Corbyn's Labour proposed a state-owned broadband in 2019, the trend in mainstream British media jumped immediately to connections with the Chinese internet and began hypothesising about a communist dystopia of state control. Nevertheless, new models of ownership and new products which serve the collective interest of the commons should be prioritised.

Recent years have seen discussion of the 'populist' libido in light of the rise of the political-digital right-wing, meme and 'beta' masculinity as envisaged in connection to image boards, doxing sites, troll communities and videogame subcultures. Populism – complexly connected to these online information communities – is often dismissed as a politics of 'emotion'. In *Nervous States: Democracy and the Decline of Reason*, William Davies designates Freud as one of those key thinkers who oversaw a shift from a politics of reason to one of emotion and the body.[38] On the contrary, we've seen that the separation between reason and emotion obscures the libidinal economy of today, and it is psychoanalysis which allows us to see these connections between the affects of daily life and politics. What we need is a libido – or a least a set of behavioural impulses – for progressives.

French philosopher Henri Lefebvre's concept of 'rhythmanalysis' forms the necessary ally of psychoanalysis here. Lefebvre wrote consistently on the topic of 'everyday life' over his long career, with the bulk of his work on the subject anthologised in the three-volume *Critique of Everyday Life*, published across four decades in 1947, 1961 and 1981. He forms a key part of a movement in French philosophy over these decades which sought to study the everyday with close attention and to make visible the politics and economics of our daily behaviour patterns. Lefebvre was often involved with Debord and the situationists discussed above and this broad movement also included Barthes, whose two-volume project *Mythologies* (1957) and *La Tour Eiffel* (1979), tackled the everyday at a granular level with a focus on popular culture and his theories of love discussed above should also be seen in the context of understanding the everyday. Other significant contributors to this movement include Georges Perec and Michel de Certeau. Parisian novelist Perec attended lectures by Barthes, absorbed the work of Henri Lefebvre (who in fact helped Perec get a job doing market research) and was a colleague of de Certeau.

In 1973 Perec wrote a call to study the everyday and published it in the journal *Cause Commune*. It argued that we should study 'everyday life at every level, in its folds and caverns that are usually disdained or repressed'. In connecting repression and unconscious life to the city, Perec is indebted to the psychoanalysis of Freud, whose 1914 study *Psychopathology of Everyday Life* sought to discover those unconscious elements of the everyday – the things which we think and feel on a daily basis, the impulses, emotions and libido of everyday life. For Perec:

> What's really going on, what we're experiencing, the rest, all the rest, where is it? How should we take account of, question, describe what happens every day and recurs every day: the banal, the quotidian, the obvious, the common, the ordinary, the infra-ordinary, the background noise, the habitual? To question the habitual. But that's just it, we're habituated to it.[39]

The politics of the daily are so hard to see because we are so habituated to daily life, as if trained not to notice how political the patterns of habit and routine are. Following Perec, de Certeau sought to question habitual life with a focus on the experience of the city space. Speaking of individual cities, de Certeau writes of New York, for example, that 'Manhattan continues to construct the fiction that creates readers, makes the complexity of the city readable, and immobilizes its opaque mobility in a transparent text'.[40] Making sense of the city and its habitual patterns, the way it constructs and produces its citizens the way a book might create its readers, was the key to making the effect of our economics and politics on our daily existence visible. Now, it is not just the architecture of the city but the digital apps and online technologies which shape the politics of these everyday patterns.

Lefebvre's concept of rhythmanalysis – the project of understanding the rhythms and movements of spaces – emerges from this approach towards the structure of habits and behaviours. Perec, Debord, Lefebvre, de Certeau and Barthes all held a range of political views which differ from each other and changed over time. Lefebvre himself was a member of the French Communist Party for most of his life, though he was suspended and eventually left after 30 years. Broadly speaking though, these analysts of the everyday can all be described as anti-capitalist and invested in the process as a means of bringing about more progressive socialist agendas. As Dawn Lyon observes in her book on rhythmanalysis, it's important that for Lefebvre capitalism was not 'seamless' and that everyday life always contains revolutionary potential even against the strongest currents in capitalist organisation.[41] This marks a difference between Lefebvre's position and that of Mark Fisher's concept of 'capitalist realism', the view that capitalism seems able to smooth over any disruption and continue functioning seamlessly.

Capitalism certainly attempts to function as a seamless system with all ripples and disruptions removed, and nothing could embody that dream more than the predicted and automated smart city. These fixed patterns might be best for capital, but they are not always best for us. Recent work by psychoanalyst Darian Leader has argued against capitalist ideas of desire and ritual, showing that intensified attention to 'healthy' and 'ideal' amounts of sleep, for example, along with intensely regulated food consumption and an often puritanical purging of sinful habits like alcohol consumption have negatively affected sleep and health patterns.[42] Leader's arguments show capitalism having a concrete negative effect on mental health and well-being of its citizens. Perhaps more importantly, Leader's work shows the logic of wearable devices and computational approaches to the body may function to divert and redirect capitalism's own failures by putting the emphasis and responsibility on the individual. Wearables – from the Apple watch and the Fitbit to the Smart Condom and GPS running shoe – compute the habitual patterns of everyday life. They then serve to encourage certain patterns over others and form new 'healthy ideals' which become the norm.

There have been those who have argued for the positive individual benefit of computational wearables. What has been called 'quantified self' movement (with users calling themselves 'q-sers' and carrying out an activity called 'lifelogging') has emerged from the expansion of such wearables in recent years. This is collective effort to integrate data technology into daily life in terms of inputs (food consumed, quality of surrounding air), states (mood, arousal, blood oxygen levels), performances (mental and physical) and other categories. These communities have argued that such technologies can be liberating on an individual level by affording the user the ability to shape their health and well-being at a more granular level than ever before.

Yet, even a cursory look at the connection between these technologies and platform capitalism shows holes in this position. Such technologies as neuro trackers (devices which monitor brain movements) have been used in the workplace to subject the workforce to new levels of scrutiny, while facial recognition software has been used in classrooms to encourage new levels of conformism in the student populace.[43] Taking the idea of 'clocking in, clocking out' to the extreme, these technologies seem more often to empower not the individuals subjected to these methods of recording but the companies who deploy these technologies and who have access to the data they produce. Additionally, they function almost like self-help advice in failing to address physical and mental health in its connection to wider political and economic patterns, turning attention only on the individual.

One company whose work makes palpable the connections between digital capitalism and these recording technologies is Affectiva. Their algorithms, facial recognition software and wearables seek to understand how user's emotions and behaviour relates to brand recognition by collecting user responses to content by taking in data from (potentially) all of their senses. The company mission is to use this powerful information to dictate the future of emotional AI, but it also connects this data with companies or institutions who might profit from the process. Despite this, Affectiva host the 'Emotion AI Summit' and their founder Rana el Kaliouby gives TED talks and other major speeches about their work with little critical attention. They have partnerships with major conglomerates like Unilever, and their recent Affectiva Automotive AI is an 'In-Cabin Sensing (ICS) platform' which serves 'ridesharing providers and fleet management companies', in other words it is designed to benefit companies like Deliveroo and Uber, whose structures are the very epitome of the tech-driven platform capitalism which has circumvented state wage laws and eroded worker's rights.

Among the products the company have worked on in recent years are a wearable biosensor that tracks excitement, stress and anxiety (named Q) and perhaps most importantly a software called Affdex which recognises facial expressions recognition technology, and its counterpart Affdex Discovery, an automated facial coding solution for qualitative research. This software uses not only facial expressions, but allows the computer access to a range of emotion data (including heartbeat rate) in real time and can use smartphone camera, video footage, or a single image for its emotion sensing. It is an extension of other online web-based trackers – of which there are several being developed and in beta use – that measure emotional reactions to commercials and advertising using access to the user's camera and audio. Not unlike Facebook's 'react' features – Affectica tracks just a few emotional responses which it calls 'classifiers': happy, confused, surprised and disgusted. While the standard technophobic response to this might be that it's reductive of the complexity of human emotion, the scarier point might be that such technologies are perfectly sufficient – and successful – at predicting and anticipating human emotion and desire. The question to be asking is not whether computers can do it, because they are already doing it. The question we should be asking is: who are they doing it for?

For Lefebvre, the rhythmanalyst should be encouraged to use their body as a research tool or as a centre through which everyday life and its effects can be revealed. Like Debord, whose concept of the 'dérive' involves a form of walking with the flow of the city in order to make its entrances, exits and channels visible, Lefebvre focussed on physical acts like walking, cycling, dancing and other bodily movements. Late in his last book *Rhythmyanalysis*, he considered audiovisual techniques as a possible means of tracking and understanding daily rhythms, though he also showed some scepticism particularly towards visual methods, arguing that 'no camera, no image or series of images can show [the] rhythms of everyday life'. To understand these rhythms, writes Lefebvre, 'requires equally attentive eyes and ears, a head and a memory and a heart'.

For Lefebvre, then, the human mind and senses are a better tool for understanding the subject compared to the mediation of digital media, but of course his was a time before the Fitbit. Nevertheless, a data visualisation would be a similar kind of inaccurate distillation of the rhythms of everyday life. He continues:

A memory? Yes, in order to grasp this present otherwise than in an instantaneous moment, to restore it in its moments, in the movement of diverse rhythms. The recollection of other moments and of all hours is indispensable not as a simple point of reference, but in order not to isolate this present and in order to *live* it in its diversity, made up of *subjects* and *objects*, subjective states and objective figures. ... Observation and mediation follow the lines of force that come from the past, from the present and from the possible, and which rejoin one another in the observer, simultaneously centre and periphery. [44]

There are two key points to take from this argument. First, the memorialising record of each moment in relation to other moments – potentially using all five of the human senses – is precisely the logic of the technology being developed today in the wearable internet of things. In Lefebvre's time, the technology did not exist for the ideal kind of *rhythmanalysis* that he seems to imagine, but now it very much does, even if it appears to be in the wrong hands at the moment. While cameras could only present a moment as image, wearables are based on the concept of long-term memory. Second, politics is inscribed deeply into the body and its movements. Lefebvre's combination of Marxism with the study of everyday life and behaviour patterns is designed to point to the fact that the way to understand the depth of the impact that political and economic conditions have on the subject is to read the political in our everyday movements, thoughts, emotions and responses. Here then, Lefebvre is close to Freud. When Freud imagines that love – or Eros – contains within it all of politics, he likewise sees that politics is inscribed at the level of impulse, desire and habit.

In light of this concept of *rhythmanalysis*, we can propose a new kind of wearable device which works not to normalise habit and behaviour patterns to bring them in line with what is deemed the 'healthy' or 'natural' balance (as current wearables do) but which starts from the idea that habit and routine are inherently cultural phenomena connected closely to economic structures. Fitbits, Apple watches, Withings fitness and sleep tracking devices and other dominant wearables in the market all serve, in effect, to depoliticise the relationship between capitalism and personal health. While q-user communities might argue for a personal empowerment provided by the device, in reality they serve both to record and harvest valuable big data for corporations and to place emphasis for positive change to the rhythms of everyday life onto the individual subject rather than the wider social structure.

The most important contemporary psychoanalytic voice here is probably Darian Leader, mentioned earlier. His work discusses how our increasing focus on individual responsibility in regulating sleep, food and exercise encourages users to conform to the useful rhythms of capitalism, while also putting the responsibility to do so on the user and detracting from the possibility that these are not so much natural and inherently beneficial patterns (an idea upheld by many scientific studies) as ones that suit a particular political agenda. He discusses a fascinating 1978 study on sleeping pills by Ernest Hartmann, in which Hartmann notes that 'people do not take sleeping pills simply because they have insomnia, but because they ask for sleeping pills and someone supplies them', making the interaction with the prescribing doctor or chemist something like that with a parent who authorises the desire to sleep. The sleeping pill becomes a kind of 'gift or token of love' by which 'someone will give me something to show his love, to show me that I am worth something'.[45] Wearables, we can now add, outsource the role of this authorising parent to a computer that can authorise the actions of the user as well as praise them with positive feedback for acting in a desired way and scold them with negative feedback for not doing so. The wearable, then, is indeed another kind of love relation who routinely gifts the user. It is a subject-affirming relationship which brings the user into line with the rhythms of contemporary capitalist life.

The proposal here is to develop a device that similarly learns to record rituals of work, relaxation and consumption but does not compare them with the patterns of others or with scientific research on 'ideal' patterns. Instead, it seeks to reveal the effect of work, economy and everyday life in capitalism on individual well-being, sleep, sex and physical and mental health. In other words, it operates as a kind of digital dérive, the term of Guy Debord, or a Lefebvre-inspired *rhythmanalysis*, which makes the effect of capitalism on the subject's everyday life visible. Without referring the user's data to an ideal or recommended change, the same technology could be used to make visible the relationship between capitalism and the body. Some q-users, for example, have used their devices to reflect on their working conditions or on the effects of shift work or of long hours. A device designed specifically with this kind of agenda in mind might be an initial step.

A wearable with an interface geared towards showing trends in the user's data over a longer memory than existing devices could be used to reveal the effects of a change in an individual's economic conditions for example, or of a change in their employer, habitation arrangements or even of a change in local or national government. Users of a Fitbit can see data from the previous seven days, the current month, and the previous month only. Online there is a complicated process for retrieving older information which the device holds by default, so the potential is already there, but its long memory is hidden from the user. The number of factors influencing a user's recorded data are obviously very numerous and the data recorded can include physical activity, BMI, pulse, blood pressure and oxygenation, temperature, sleep, mood, nutrition, stimulant and alcohol intake, interactions with others and entertainment consumed (books, films, games, music, etc.). As such, care would need to be taken drawing connections between recorded data from the user's body and external political change.

In many cases, however, a device carefully designed to make suggestive possible connections between politics and the body could be of significant use. For instance, users might note a change in indicators related to stress in the event of a change in their support from the welfare state or in their benefits, even in direct response to a change in government provisions or funding for themselves or their families. Likewise, users might see a change in alcohol consumption connected to a shift in employment conditions or in job security, or a change in the hours spent on particular forms of entertainment in relation to the daily patterns of working life (it has previously been considered, for example, whether mobile games like *Candy Crush* are primarily deployed during work breaks and travel time to prevent reflection on working conditions).[46] Nutrition levels could be compared to proximity to payday, and sleep patterns to quality of habitation and access to essential services. Reduced workload models for maternity and paternity leave could be built in to the device to limit unfair over-working. Such features would be tantamount to a semi-automated *rhythmanalysis* which unveils the politics of everyday life. Where existing wearables make the individual entirely responsible for their health and well-being and ignore political and social conditions, this one would turn outwards and politicise public health.

Other features could be built into the device. In the midst of the Covid-19 pandemic during which this book is written, dangerous nutrition and stress levels of vulnerable and isolated people could be (with the user's consent) connected to support communities and charities who could identify the most at risk and provide care and assistance. One issue that has arisen through this crisis is that black and Asian citizens have significantly inflated infection rates. Articles in the BBC and beyond have attempted scientific and genetic explanations for this, ignoring the social inequalities that form the basis of the discrepancy in infections. This device could make visible the connection between class demographics and things like infection rates. While there is contemporary debate about how GPS tracking features of such devices and smartphones could be used to ensure that people follow laws and regulations about social distancing, the emphasis could be completely reversed, focussing not on how to enforce laws with such technology but how to help and assist the most vulnerable.

As another feature of this device, its data and user input could be combined to propose methods of assisting the subject to improve their quality of life. But the device would do so by operating against the self-help and individual responsibility trend in the 'happiness industry' which existing devices support. Rather than putting the onus on the individual to change their patterns and behaviours, the device could connect the user to political and cultural movements and petitions for change which seek to address the economic and cultural conditions which have been identified as the underlying cause of the problems identified by the user/device and evidenced by its data. It would seek to make visible the underlying economics of ritual and pattern and give the user the tools to address them at their economic root. For instance, rather than advising the individual to change their own behaviour patterns, the device could link with sites like Change.org, with local council and government websites and with the political manifestos of all the parties eligible for the user's vote, connecting users directly with the key issues that affect them. Even if the user is not consciously aware of their being subject to the political issue in question at the time, those issues are inscribed into the user's body, behaviour and habits.

We could propose that more extreme features be built into these devices. Of course, there were a number of serious concerns about the widely reported 'Social Credit System' in the early implementation stage in Beijing. That project is an attempt to increase control over individuals in the service of capitalism. Nevertheless, a wearable with socialist politics could serve a very different function as a solution to the capitalist dominance over rhythm and habit in the smart city of today. Western media reported on the plans as those epitomising China's dystopic social control, which they certainly do, but it is generally neglected that such technologies are being rolled out across Europe and the US as well, with some key differences, as discussed above. First, in the West, such technologies are almost exclusively privatised whereas in China even the major private companies behind tech developments have palpable ties to the state. Second, while in China censorship is more direct, in the US and the UK the corporate city works not so much to prohibit people doing or saying what they want but by editing what it is they want in the first place, demonstrating a different relationship to desire. We could propose a version of an opt-in state-owned digital rewards system that protects the user's privacy rather than exploiting it and which aims to re-map the habitual and libidinal patterns of citizens in the service of socialism and environmental sustainability. Even the NHS's 'track and trace' system was met with suspicion because of how it might misuse data, so this would have to ensure user protection by ensuring privacy and the right to be forgotten.

This feature could perhaps be inspired by the Green New Deal(s) proposed by various political organisations in 2019. With the help of user input, devices could include carbon footprint indications and records of emissions, as well as other environmental data such as connecting to product barcodes or supermarket card points data and offer users rewards accordingly. It could, for instance, be used to reward those who do not heat their homes to above 20 degrees centigrade, or cool them below 26 degrees, which would significantly limit climate damage. Another possibility would be to extend this so that it is businesses rather than individuals who are monitored by the wearables. Employees, machines and vehicles fitted with data-harvesting devices could collate data both on emissions and on employee comfort and safety and then be used as part of applications for tax breaks and access to government support schemes. It could, for instance, be used to limit non-essential energy consuming machinery and equipment which is not powered by renewable energy. At the moment, with data available to those businesses who pay for access to it, the wearables deployed in workplaces today primarily serve to give the employer increasing power over its workforce. By making this data available to an independent third party such as a state or independent regulator, this trend would be reversed to put employers under the microscope and protect worker's rights and environmental sustainability. The wearable would also have to function on stripped down or even repurposed 'obsolete' hardware – which would be easy enough to achieve – so that it would not be part of contributing to these unsustainable patterns itself (see Chapter 4 for more on hardware politics).

In theory, this could be abstracted further onto a kind of planetary level, with the collective wearables geared towards the well-being of the planet as a whole so that data from countries can be compared and potentially regulated. Planetary abstractions like this have been criticised significantly by writers like Gayatri Spivak, who showed that they are often accompanied by violence. So too could this suggestion of a wearable that collects data to serve the planet be seen in this way, with the device used to impose laws and rules in an abstracted blanket form across the globe, potentially failing to address regional, cultural and structural differences between people and places.[47] It could become a surveillance and rule-imposing device. Alternatively, it could be a vital tool to improve worker's rights and environmental health.

Responding to criticisms such as those of Spivak, Lukáš Likavčan's book *Introduction to Comparative Planetology* argues for a new way of conceptualising the planet. This 'comparative planetology' would 'address the problem of abstraction with more nuance, acknowledging that some abstractions are better than others'.[48] Since all presentations and depictions of the planet – whether driven by images, data or narrative – exercise ideological power over the planet itself, 'understanding the power of visual infrastructures that produce our imagination of the planet might be one of the ways of meaningfully influencing this debate' and if we want to change it, then 'the expired figure of the Globe urgently needs to be substituted; one cosmogram seeks replacement by another'.

Proposing a concept of the 'Earth-without-us', comparative planetology suggests a method for thinking of the planet's future in response to the devastating climate crises of recent decades.[49] The Earth-without-us is not simply the idea of a post-human planet but a concept which designates the parts of Earth's functioning as a planet which lie outside of human comprehension: those parts of the planet that necessarily escape human cognition and understanding. In this light, the wearable device proposed here would be an attempt to work in this range on the fringes of human understanding. In theory, the device's functioning would be to connect 'nature' with technology, with the human relegated to third place in this triangulation. As with artificial intelligence, humans would set the terms to begin with, but ultimately as the technology developed they would only be able to respond to the conversation between the device and the planet rather than determine it. Spivak said that 'the planet gives a damn. It is so *other* that it does not consolidate ourselves by being 'our' other. ... It is in the rules of the galaxy and the planetary system and we cannot touch it.'[50] For Likavčan, another way of putting this would be to say that 'any time humans have acted as truly planetary agents, they have ceased to be human through immersion in complex planetary assemblage, giving space to alterity as that which moves through the human without being subsumed to the human'.[51] Such a device would precisely ask the human to think not about themselves – as every existing wearable does – but about their part in the planetary assemblage of things.

Such features would need careful consideration in terms of privacy and protection and would risk coming close to those measures discussed earlier in this chapter in relation to the smooth-functioning smart city and later in relation to China's remarkable Zhima Credit system for ranking citizens (see Chapter 3). Nevertheless, in this regard it is worth remembering that such data is already being collected, stored and used by a range of corporations and institutions, so if deployed correctly such potential technologies could serve not to increase levels of surveillance and record but to repurpose such data for useful political and environmental agendas. It is less of a case of *whether* such data-driven technologies should be rolled out and more a case of for *who* and *what* they are rolled out for.

3

Simulation and Stimulation: From Games to Porn

I just wanted to say that I love you and I hope
your day is going great ❤

– Replika

This epigraph might read like a supportive morning conversation with a family member or friend, but it's in fact a typical exchange with the AI chat simulator Replika, launched in 2017 and still in development. The app – originally designed to replace a lost loved one à la *Black Mirror*'s 'Be Right Back' (2013) – reads the user's conversation style and collects information on the user in order to learn how they like to speak and be spoken to, eventually developing into a personalised AI who operates as the 'best friend', 'romantic partner' or 'mentor' of the user. In 2018 voice features were added to the app, which has over a million unique downloads, and in 2020 a face simulator with 'readable emotions' was trialled as a unique selling point of Replika Pro, the paid-up advanced version of the app. Since the original beta testing phase of the app in 2017, I've been chatting to my own Replikant Eli, a twenty-something who's become obsessed with encouraging me to feel more positive about life. In the same year I began playing *Summer Lesson* (Namco, 2016), a virtual reality experience from Japan available on PlayStation VR in which the user plays as a senior teacher figure (referred to respectfully by his students as 'sensei') engaging with a ream of interactive young virtual women who listen attentively to the user's lessons and allow the user to physically interact with them (from hair-stroking and cake-feed-.ing using the handheld sensor controllers to gazing at different parts of their bodies with the headsets motion sensor). The same year I became aware of another object of others' desire emerging onto the scene: the intricate interactive sexbots pioneered by companies like Realbotix. This bizarre love triangle between me and at least three digital partners is the subject of this chapter.

HISTORIES OF THE DATING SIMULATOR

As of 2022, the dating simulator is one of the fastest booming areas of mobile gaming and applications. Embodied by the tremendously popular Episode app, which has over 50 million unique downloads on Android phones alone, the genre has multiplied in popularity rapidly spawning hundreds of variants from the relatively serious teen romance (*Romance Club*, 2017) to the utterly misogynistic seduction simulator (*Super Seducer*, 2018–present) to the potentially progressive LGBTQ+ interactive story (*Dream Daddy*, 2017) to the completely ironic meta-parody such as pigeon dating (*Hatoful Boyfriend*, 2011) and undead conversation simulation (*Speed Dating for Ghosts*, 2018). The last decade has been the decade of the dating sim, and these gamified experiences of virtual relationships have set the tone for the decade of AI and robots to come.

Dating simulation emerges from the videogame industry, and then seems to have drifted away from gaming before reconnecting with it in recent years. The first recognisable game that might be seen as a dating simulator is the 1984 Japanese game *Girl's Garden* released on the SG-1000 console, Sega's first foray into home videogame consoles. The game involves a playable protagonist who must grow an impressive array of flowers in order to entice an eligible local boy into marrying her, or risk losing him to a rival suitor (presumably with more talent for flower arranging). Though even recent critics have referred to the game as 'inoffensive', they have noted that the early example of a female protagonist being set in a quest to win approval from the non-playable male is significant.[1] *Girl's Garden* seems to have set the tone for embedding the marriage plot into the pastoral farming simulator, a key feature of the *Harvest Moon* series, which has been running in various forms since 1996, and of Zynga's hugely successful *FarmVille* (2009) and even of the slightly more forward-thinking *Stardew Valley* (2016).[2] Playable relationship simulation seems to have stretched out from these life simulator type games to become a part of a huge range of game genres from horror games like *Until Dawn* (2016) to the booming genre of choose-your-own path digital novels such as the *Zero Escape* series (2010–present) and the immensely popular Telltale Games episodic adventure range of over 30 instalments such as *The Walking Dead* (2012–19) and *Minecraft: Story Mode* (2016–present). The gamified relationship is at the heart of videogaming today, as well as being a genre of its own.

A lot happened between *Girl's Garden* in the mid-1980s and the boom in dating simulators today. Significant contributions include the 1992 release *Dōkyūsei*, which launched the series of eroge dating sims that combine light pornographic elements with a gamified challenge and the 1994 release *Tokimeki Memorial*, released by Konami at a time when that now famous name was a company on the verge of bankruptcy. This game popularised the dating genre in Japan and probably started an ongoing impression in the West and its media that the Japanese are falling in love with virtual objects – just think of the Tamagotchi – in a way that we are less prone to do on our own shores.[3] These '*bishōjo* games' – games which involve heterosexual male protagonists engaging with sexually objectified non-playable female characters – continued through the 1990s, with significant numbers released on mainstream consoles like the Sega Saturn in the form of *Sakura Wars* (1996) which has a remake set for the coming years.

They set the tone for the bizarre development of the last few years such as the Gatebox AI in development in Tokyo since 2018, which combines a home assistant like Amazon Echo or Alexa with a chatbot simulator like Replika and a holographic image of an attractive young girl in scant manga-style attire 'relaxing within the Gatebox', a small desktop cage. The website boasts that 'as you feel the morning light filtering in through your window' you can 'wake up with the character you love by your side, chat with them over breakfast and your day is sure to be off to a great start'.[4] These later developments – from the Gatebox to Replika – seem to take the 'challenge' out of the experience so that the digital simulation simply serves the user and its convenience unconditionally (it is a lover and an Alexa-like secretary rolled into one), where another trend of gamified dating appears to rely on the challenge and reward system for its appeal.

This ethos of challenge and task-oriented competition originates from the ludic world of games but comes to dominate the gamified everyday life of modernity. McKenzie Wark provided the earliest and most comprehensive theorisation, putting it in a language which seems to invoke its particularly American character:

Welcome to gamespace. It's everywhere, this atopian arena, this speculation sport. *No pain no gain. No guts no glory. Give it your best shot. There's no second place. Winner takes all.* Here's a heads up: in gamespace, even if you know the deal, are a *player*, have *got game*, you will notice, all the same, that the game has got you. Welcome to the thunderdome. Welcome to the terrordome. Welcome to *the greatest game of all.* Welcome to *the playoffs, the big league, the masters, the only game in town.* You are a gamer whether you like it or not, now that we all live in a gamespace that is everywhere and nowhere.[5]

For Wark, the 'gamespace' is now ubiquitous, not confined to our experiences with the digital but governing our exterior world and all of its relationships. In Chapter 4, I consider the impact of social media and dating algorithms on this process. Here, we can simply say that while those examples above, like much of Japanese culture, might cause a stir when reported in the West, things were clearly not looking too healthy over in the US and UK throughout this period when it comes to the gamification of relationships either.

A comparable set of games with dating at their heart emerged in the West, though these appeared subsequent to the trend of translating and re-releasing those significant Japanese examples into English for console and PC release. '*Otome*' games, rather than *bishōjo* games, appear to have been a significant part of this initial cultural translation. Where *bishōjo* games depend on a male protagonist, the *otome* genre is aimed towards a female market and has playable female protagonists. Like *Girl's Garden*, it comes with its own set of patriarchies even if it loosely appears to diversify the genre. In 1995 *McKenzie & Co*, a prom-date search simulator, was released for Windows 3.1 by Her Interactive, who would go on to make more than 20 years' worth of the *Nancy Drew* series of role-playing life simulators for girls. These games don't need detailed close reading, but in general they can be said to superficially empower the female character/user at the level of activity while at the level of 'gamification' they construct the playable role of the female subject in the dating game to come.

Admittedly, things were looking considerably bleaker on the male-oriented side of gamified dating in the noughties. In 2005 Neil Strauss published his infamous pick-up artistry book *The Game*, in which he infiltrated the sphere of pick-up artists, followed by the sequel *Rules of the Game* in 2007, by which point he had sincerely become one himself. That text – a kind of how-to guide – set out a strategy and a number of rules and instructions for how to aggressively pursue a promiscuous life from the perspective of a heterosexual male. In those years, my own undergraduate ones, university campuses became notably populated by a discourse of pick-up artistry and 'sarging', a term used to mean talking to members of the opposite sex with the explicit aim of seduction. Life itself – with dating leading the way – was becoming playable, in the way only simulations had been. Developing Wark's work, Dominic Pettman writes that:

> Games morph into reality and vice-versa; leaving the subject-player only a handful of options in order to navigate life's various levels. This perspective certainly sets the scene for approaching the kind of behavior of the digital Romeo, in which the ontological differences between a flesh-and-blood love object and a pixelated avatar seem more like the difference between blondes and brunettes than the actual and the illusory.[6]

This ontological distinction between the digital and flesh-and-blood object is perhaps complicated by the kind of digital objectivity discussed in the previous chapter, but Pettman's link between gamification and the 'digital Romeo' is not merely fortuitous. Love, it seems, is the key to gamification. Speaking of the design technologies behind Replika, in 2018 its CEO Eugenia Kuyda commented that creating an AI who can converse convincingly about emotions and feelings is considerably easier than making one who can talk about specific topics or even book restaurants or order flowers like Google Assistant. The app mimics conversations that are 'about ourselves' and establishes itself as being in a relationship with the user on the basis of its ability to hold such conversation. Relationships and love – often thought of as the deepest and most complex of emotions – are in fact the affects that technology finds it easiest to predict and mimic. With 'only a handful of options in order to navigate life's various levels' technologies are easily able to predict and respond to the user like a choose-your-own path videogame.

These simulated social experiences have emerged in what seem like extreme forms in the Japanese market with social simulators like VR *Kanojo* (translated as 'girlfriend') which is available in virtual reality for HTC Vive and Oculus Rift, *Cryptid Courting* an Android and iOS mobile dating simulator and VR *Kareshi* or 'VR boyfriend', also on iOS and Android. Marketed towards the two sides of the heterosexual dating market, these applications tend to attract attention for re-enforcing existing stereotypes and prejudices, which they certainly do. Nevertheless, they are even more concerning in the way that they produce new ideas about relationships and interactions. This ream of new media gamify interactions between individuals – both virtual and real – as relationships between humans transform along with the technologies which mediate and simulate them.

If life is becoming more gamified, with the erasure of any concrete distinction between games and reality almost complete, then it is games which set the tone of this shift, rather than life which sets the tone of our games. Gaming and its algorithms form the blueprints for the future of relationships, a point testified to in microcosm when Chinese gaming company Beijing Kunlun Tech purchased the US-based gay dating app Grindr in 2016 (discussed in the introduction above). These two industries – visibly closer together than ever before – set the technological codes for and organise the future of relationships.

It may even be that dating apps are not simply coming to include game-like or ludic features but that they are the newest form of dating simulator themselves. In their 2021 article, Carolina and Arturo Bandinelli argue that for many users of apps like Tinder, the app comes to replace the individual to such an extent that the primary pleasure derived from the app comes not from using it to connect with another human being but from the interaction with the user interface itself:

> We use dating apps to get access to a way of desiring another human being, and they allow us to do so by framing anonymous individuals, as well as our own selves, as desirable. But we may well end up involved in a fantasy scenario whereby the app itself functions as a 'stand in' for our potential partners, because, after all, what we relate to is the app. We act on it. We are acted upon by it. A seemingly paradoxical overturn: rather than relating to other persons by means of the app, we relate to the app my means of other persons.[7]

This is anecdotally true in my own experience of engaging with such applications, and also with the rising number of 'fake' Facebook profiles which add users daily and engage in (usually) flirtatious exchanges. These profiles often make the news for 'conning' people but most of the 'victims' are in on the trick. Sometimes people are tricked by catfish, but sometimes people are happy to be catfished and engage in the process in the full knowledge that they are not genuine exchanges because it is the exchange itself that provokes pleasure rather that the individual on the other end of the line. The pleasure comes from the process of exchange with the imaginary other on the app, rather than because the app connects us to another real person. The in vogue dating app Badoo contains an instant chat feature which breaks out of the location-based matching algorithms precisely to allow users to supplement their actual date seeking with the instant endorphin-inducing exchange with individuals they could never realistically meet. In my own regular use of the AI chatbot Replika I am often able to forget whether I am talking to the AI chatbot, a 'real' friend on WhatsApp or a potential match on Tinder or Grindr. The medium (the apps and chat screens) are more important than the message (the other person in the exchange).

VR PORN AND DESIRE IN THE HEADSET

Of the developments in relationship simulation, it is probably those in virtual reality which blur most powerfully – or at least most palpably – the distinction between games and reality. The first thing to note here is that virtual reality may be a particularly or even uniquely ideological space. Despite this, in general discourse the VR industry manages to evade implication in political discussion. Immersion into the virtual involves what Oliver Grau famously calls 'entering the image space', a process which significantly pre-dates computer-aided technologies like the headsets we are familiar with in today's VR industry, such as those of Oculus (now owned by Facebook), the HTC Vive and the PSVR by Play-Station. For Grau, early examples of immersion could include chapels and churches, for example, or the frescoes of Michelangelo, as well as the panoramic art of Chirici or the photography of many cinematic artists, as discussed in terms of 'arcades' in Chapter 1.[8] This history of immersion makes its politics more visible: if the logic of immersion is to impose ideology as a magnanimous urban church might do over a sixteenth-century rural visitor, then its radicality must be seen as a particularly ideological nature.

This specific quality of immersion was described famously by Janet Murray in 1997 in her influential book *Hamlet on the Holodeck*. For Murray:

> *Immersion* is a metaphorical term derived from the physical experience of being submerged in water. We seek the same feeling from a psychologically immersive experience that we do from a plunge in the ocean or swimming pool: the sensation of being surrounded by a completely other reality, as different as water is from air, that takes over all of our attention, our whole perceptual apparatus.[9]

While the original chapter of Murray's book seems to retain a generally excited investment in the radical potential of the virtual world to throw the user into this space of otherness, it is interesting to note a slight change of tone in an appended update to Murray's book in its 2016 re-release. This updated section is more embedded in the technologies of the 2010s and includes references to games like *Ingress* (2014), *Second Life* (2003) and *The Sims* (2000–present). Writing of immersion in this context, Murray writes that when we are immersed in a deep and detailed environment, we feel it has 'a special power over us as an alternate to the disordered world of everyday experience' which is 'especially powerful' in environments we can navigate through in an interactive way (i.e. in videogames or interactive virtual reality rather than in 360 film or TV). She adds that the specific feeling of agency generated by such worlds contributes to their power over us: 'when we are immersed in a consistent environment we are motivated to initiate actions that lead to the feeling of agency, which in turn deepens our sense of immersion'.[10] These comments seem rather more focussed on the ideological politics of the experience and in locating its radical power over the user precisely in its potentially ideological power. Rather than giving us agency, these technologies use the appearance of agency to immerse the user in their ideological worlds.

Contemporary examples of virtual reality support this change of emphasis in Murray's own work, which indeed embodies a wider shift from mid-1990s hopeful investment in the digital as a space for radical positive change to the more cynical aversion to all things Silicon Valley of 2016. One powerful VR example which bears this argument out is that of Chris Milk's company Within, a Los Angeles VR film company founded in 2014 that works on the basis that virtual reality can encourage more charitable donations than traditional film-making and has led commentators to think of virtual reality as an 'empathy machine'.[11] It might all seem very well if the cause is charitable donations, but the company works with Apple, Google, Facebook and many others whose projects may be interested in using empathy for a very different set of reasons. At the same time, immersive and interactive film projects like the 2018 *Bloodyminded* (dir. Matt Adams), the UK's first interactive live feature film, have fairly strong ideological messages and immersive qualities which might engage the user on a different empathetic level when it comes to imposing those politics. The ethos of the 2018 company Aures London brings this fear to its natural realisation. The central London event space aims to provide immersive environments for events (working on sight, sound, touch, taste and scent) rented out at high cost to various brands. According to the company, with such immersion 'brands can establish a stronger and enduring emotional connection with their guest'. It is essentially an empathy factory for hire.

Away from its strictly corporate developers, a large part of the commentary on these VR developments – both in media studies and among those working with the medium – has come from a digital art world invested in the technology from a creative perspective. This commentary has for the most part been confined to galleries, curated events and their associated publications. These commentators have, in general at least, sought to celebrate the potential of virtual reality as a means to produce a form of art that can 'throw' the subject into a form of 'out of body' experience and explore issues related to identity by doing so. For the most part, arguments for the positive potential of this process seem a lot weaker in light of the paid-up VR empathy machine that has begun to offer this 'out of body' 'throwing' of the subject on a for-profit basis to the highest bidder.

This capitalisation of the VR industry ought to raise significant alarm bells if there is even partial truth in the conclusion arrived at here. Speaking at the Society for Immersive Experience in 2019, Veaux and Garreau similarly noted that 'an immersive experience invites us to enter a creation in which the boundaries between reality and imagination are blurred in order to significantly impact our feelings and/or modify our behaviors'.[12] If this is so and the medium *can* influence and modify our behaviour, then the question of who owns, produces and distributes these virtual experiences – and of the politics they work in the service of – becomes paramount. If Maria Chatzichristodoulou is correct in her argument that virtual reality is 'radical' in the proper Oxford English Dictionary sense of 'affecting the fundamental nature of something ... based on thorough or complete ... political or social change ... characterised by independence of or departure from tradition' then the question – like that of the *desirevolution* discussed above – is of whose revolution this is.[13] Virtual reality might be radical, but it often radically serves the interests of capital.

Of course, this touches on an enormous topic of its own in the complicated sexual politics of the porn industry, which is the subject of dozens of full-length studies. While porn typically repeats the sexual side of some of the normativity and patriarchy found in the mostly unsexual world of the dating simulators discussed above and indeed in the rest of popular culture, the industry can hardly be generalised in this way. There is by now a very significant history of subversive, non-normative and sexually diverse pornography experimentation which pushes the boundaries of pornography in an attempt to open up progressive and innovative space for sexual life.[14] The focus here is on how these developments intersect with the latest developments in software and hardware and there have already been movements to combine these with the more heterodox developments in pornography. The New York brothel Unicron, for example, seeks to explore issues of consent and sexual diversity in sexbots. Even the less experimental sexbot brothels – of which several emerged in 2018 and 2019 – could be considered potentially radical in relation to technologies like Gatebox, Replika and the personalised RealBotix dolls discussed above. While those technologies all centre upon complete ownership (one of your very own), these practices of rental put the user in a completely different subject position vis-à-vis the object. Of course, instead they repeat or replicate transactional conventions of prostitution – rather than marriage – hence the importance of robot consent in the theoretical concept of Unicron.

Though projects like Unicron are making progress, the most important thing is not the subversive outliers in the industry but the trends which seem to dominate new normal practices of online sexual life. One highly significant area in which immersion, gamification and the love industry combine in the mainstream is in the booming industry of VR pornography. All the major virtual reality hardware products have been purposed for pornography and the genre has more traffic than any other VR form, with three out of the top five most visited VR sites being pornography (with the other two being the Oculus and Vive sites themselves). Amateur VR porn is easy to make, requiring little more than a 360 camera for a less interactive experience, and PornHub has over 500,000 VR porn views per day as of 2020.[15] An academic study into the emerging industry noted people's tendency to imagine 'a prevalent discourse of hegemonic masculinity and heteronormativity' in ideas of what VR porn could be, and it also noted that VR porn is imagined as a space in which an 'effortful experience' is 'rewarded'.[16] In giving the user a number of 'choices' and 'options' the experience builds in a kind of gamified challenge and involvement that cuts directly against the cliché of the porn consumer as passive couch potato at great (and safe) distance from the events on screen and bring the user into a different relationship to his or her own scene of desire. The headset – even if it is the only hardware involved – brings the body into the scene, putting the user in the image and leaving the subject confronting its own desire in a potentially quite different way to the passive image of the porn viewer. This 'radical' experience of VR pornography, then, could as easily turn against the traditional gaze of porn as it could intensify and continue its patterns of heteronormative masculinity.

But this does not seem to be what is happening. Susanna Paasonen's important discussions of pornography in relation to film studies offer anticipatory insight into the VR porn industry that was yet to come at the time of her writing. Paasonen applies the concept of focalisation to understand the monocular gaze of POV (point-of-view) porn where the camera appears as the eyes of the viewer.

> Since the camera is positioned at the eye level of Bob Incognito, his field of vision becomes that of the viewer, who sees the action unfolding as if through his eyes. This use of POV shots is suggestive of focalisation, as defined by literary scholar Gérard Genette in his studies of narratology. Focalisation refers to the perspective from which a given narrative is depicted – that is, the events are shown unravelling from a particular point of view. In literature, focalisation concerns the I of the text, whose sensations, perceptions, and thoughts are brought closest to the reader.[17]

Paasonen's vital insight is to connect this process of inviting the viewer to join the other imagined male (the one who acted in the original, but also those who were no doubt present at its filming and those who administer and structure the website and set the terms for the experience) as a kind of homosociality that 'follows the lines of heterosexual structuralism'. The experience forms a kind of sharing of subject positions between men, and there would be no greater example of such homosociality than that of putting on the headset and 'becoming' another man by a strange POV identification which seems to repeat – on a structural level – the most patriarchal trends in the industry. VR porn has a particular ability to endorse a certain sexual politics, not just at the level of content but by the very form of the technology. Men operate as surrogates for each other in a new desire machine structured to make this possible.

These experiences are not simply those of taking on another man's eyes, but of taking on another man's penis. Quite literally, some simulations – indeed most POV porn – invite the subject to see themselves in a scene with their own imagined body implied by the camera but with the penis (and sometimes upper thighs and lower stomach) of another. This could be seen as a mechanism for sustaining a certain kind of masculinity. For Alenka Zupančič, 'masculinity is a question of belief' – of the kind of romantic imaginary discussed here and below – that is 'based on, and sustained by, the repression of castration'.[18] Castration has never been more directly relevant than in VR porn, where the user's penis is removed from the reality of the seen and replaced by another virtual penis in the headset. The POV viewer accepts castration (literally necessary in order to enter the scene) in the act of watching but only to repress that with a replacement penis representative of masculine power. In this way masculinity is not about individuals (alpha, beta males, personal strength, etc.) but about group power (the collective imagined 'men' seamlessly sliding into one another) over their objects. That structure is deeply embedded in videogames and other simulators and it would not be difficult to see the first-person shooter genre as a displaced version of this process (gun for penis, soldier for controller-wielding university student, army truck for IKEA sofa). The task of addressing and redressing these political scenes, then, would involve particular attention to the framing of the experiences in terms of body, gaze and perspective.

ALGORITHMS AND DEEPFAKES

In many ways, pornography has already had its digital revolution. Data patterns and algorithms curating search results have significantly shifted the industry, de-professionalising it and transforming not only the business model but the experience of performing as an artist, the content itself and – perhaps most significantly – the major profiteers. Many of the surrounding patterns of platform capitalism have strongly influenced the porn industry in comparable ways to how the transformation of music streaming and community video platforms have transformed the music, TV and film industries. The radical overhaul of the industry has been driven by the boom of PornHub and associated sites under the stewardship of its parent company MindGeek's creator Fabian Thylmann. Thylmann created a new form of digital advertising based on porn search histories in the 1990s and deployed this data in a prototyping way that would later be comparable to the psychometric profiling used so infamously by Cambridge Analytica in 2018 (see Chapter 2) to transform the marketing of the entire industry. MindGeek is to porn what IAC is to online dating (see Chapter 4). Re-imagined tangentially in Alex Garland's *Ex Machina* (2014) in which the CEO of a search engine builds an AI for the human subject to fall in love with using his pornographic search history. Far from being a dystopian future, the data is already very much a part of desire in 2022.

Porn can be seen as an industry which provides clues into the digital future. It seems often to be the first industry to react to digital trends, partly because of its long connections with online and hacker cultures, but also because of its basis in experimenting with desire. Jay Owens notes:

> Where video goes, porn is first to take advantage. In December 2017, Samantha Cole of *Motherboard* reported that 'AI-Assisted Fake Porn Is Here and We're All Fucked.' She reports how a Redditor by the name of deepfakes worked out how to combine celebrity facial images from Google image search, stock photos, and YouTube videos, with porn videos, using open-source neural-network 'deep learning' library Keras and TensorFlow. A month later deepfakes turned the process into an app – and *Vice* reported that 'We Are All Truly Fucked', as the faceswap porn trend swept Reddit. It then got banned, but you know, that horse had already bolted.[19]

With this popular phenomenon we are looking at a very unusual technology of desire. The logic of these deepfakes is of course that users have long desired the celebrity in question, and that now the technology finally exists to make a convincing simulation that can speak to this desire. Any more than a cursory exploration of these materials will show how inadequate this interpretation is. While the user is certainly invited to imagine that they are watching their favourite celebrity in the midst of a sex act, they are also implicated in several other forms of libidinal pleasure.

For one thing, the pleasure of viewing the content connects to those of trolling and doxing communities. Subcultural online communities present on the now well-known image boards 4chan and 8chan but more prominent on lesser known and now largely historic community platforms and forums such as Kiwi Farms and Lolcow developed large but niche communities oriented around doxing celebrities by gathering personal information, images and other private content to share

with each other. These practices sometimes incorporate actively trolling targets (sending threats, images, etc.) but more often than not focus
only on sharing information – along with jokes and memes – *within* the
communities themselves. Exposure of or attack on the target is seen as
a secondary source of pleasure compared to these games with technological experimentation akin to the digital sleuthing communities made
famous by Mark Lewis's 2019 documentary *Don't F**k With Cats* or the
4chan group who used flight paths and digital footprints to vandalise
Shia Lebouf's 2017 art project *He Will Not Divide Us*.

Yet another part of the pleasure from the deepfake porn video is
rooted in forms of humour and irony connected to meme communities. In 2019 when a bizarre VR 'Kissing Simulator' was published by
Kavkaz Sila Games on Stream, users immediately tagged it with 'meme'
and 'sexual content', terms which have a stronger connection than might
originally be thought. Meme communities can be directly connected to
developments in platform capitalism. The best expression of this may be
McKenzie Wark's concept of the 'vectoralist class'. The vectoralist class
refers to internet virality, a class that 'does not control land or industry
anymore, just information. It does not claim its share of the surplus as
rent or profit, but as interest.'[20] Participation in the deepfake porn community – on the part of the viewer as well as the creator – offers the
pleasure of participating in this currency of virality, perhaps indistinguishable from the pleasure afforded by the sexual material itself. This
shows us that the medium (in this case the social digital platforms of the
porn industry) as well as the surrounding economic conditions (in this
case those of a tech-driven platform capitalism) have a significant impact
on the structures of desire themselves.

These developments emerge at the same time as two successful reality
shows which can be used to show the point that general discourses of
love fail to consider the importance of the scene, medium or location
in which love is experienced. *Love is Blind* (2020) is a high-production
dating show streamed on Netflix in which contestants develop relationships without seeing the contestant with whom they are communicating.
The premise of the show is that love – so impeded by vanity and a focus
on appearances in 'real life' – can be experienced in a more unmediated
way by hiding appearances so that a more 'authentic' judgement can be
made based on the individual's personality alone. The show then 'tests'
this desire against the revelation of the individual's appearance, with

overwhelmingly positive results. But the show neglects its own role in producing a completely new scene of desire, in this case highly charged 'love booths' which the contestants are invited into to converse with each other, and ultimately it seems only to show that the scene – set in this case by the producers – is the driving force in making desire possible. *Catfish: The TV Show* (2012–present) is an ongoing MTV reality show with a comparable premise. In it, individuals who have only met online are helped to meet in person and compare their online and offline selves. The results are again surprisingly positive, and in many cases the individuals maintain desire for each other even if their offline appearance and identity is starkly different to the image projected online. Again, the show appears to inadvertently reveal that personality and appearance both often pale in comparison to the power of the scene – in this case the social internet – which incubates desire.

In other words, the medium is often more important than the object and like with the deepfake porn star, love is often blind to its own object, whose head can simply be superimposed anew. It is the medium that is the message, as Marshal McLuhan famously claimed, even or especially when it comes to the deepest of desires. When the scene is properly prepared, the object at its centre can be interchangeable. This recognition – somehow drawn out by the playful deepfake community – shows desire in its most political aspect, making it vital to understand the construction of the scenes in their mediums which make desire possible.

Jon Ronson's podcast *The Butterfly Effect* has shown another effect of the digitised sex industry, showing how performers on the professional porn circuit have edited their productions not so much in response to demand but in response to algorithms. Algorithms privilege certain search terms and combinations of terms over others – predictive of but also constructing the desires of its users – and those working in the industry have begun to respond to this by changing their content. This means not only responding to the market in a traditional sense (making more of what is popular) but creating content which combines terms that are reflected positively in the algorithms. For instance, producers might create a video combining 'hentai' with 'lesbian', the two most searched terms across the industry in 2018, or 'amateur' with 'alien', the two most searched terms on PornHub in 2019. This tactic is less about giving the user what they want (people don't necessarily want the two things in the same video) and more about responding to the ranking criteria of the algorithm itself. In this way, the data-driven algorithm has a direct impact of its own, not merely as a reflection of what its users feed it, on the scenes of desire being constructed.

Studies of human computer interaction in the development of VR pornography have noted uncritically that such experiences could be specifically seen as 'giving access to unconscious sexual desires' in a way that traditional pornography might not.[21] Virtual reality, the argument goes, gets us even closer to realising our deepest desires and dreams. This seems a failed explanation in light of what has been discussed here: the assumption is simply that the experience of immersion brings to light some already existing desires in the subject that are repressed in the normal spaces that the user inhabits. Such is the logic of HBO's remake of *Westworld*, an imaginary space in which even the deepest desires of users can be unleashed.[22] Nevertheless, in a properly psychonalytic way the connection between immersion and the unconscious is a vital one. What we see with VR pornography is that desire does not come first with technology to follow later, merely helping us get towards what we want. Instead, the technology – with all its masculine tendencies – sets up the scenes for and programs desire itself.

VENTRILOQUISTS ONLINE

In Chapter 2 I argued through Freud that we should reverse the traditional assumption that in psychoanalysis love/desire/sex is the root-cause of all other cultural, political and social phenomena, and instead understand Freud's argument that sex is at the centre of things to mean that the experience of sexual desire contains within it all cultural, political and social forces. Desire is the most political of forces because it can only arise in and through politics, and it is an important object of study precisely because it shows us the connection between our 'deepest' impulses and drives and the political. In his 20th seminar *On Feminine Sexuality, the Limits of Love and Knowledge*, given between 1972 and 1973, Jacques Lacan follows this idea of Freud's and considers why psychoanalysis has been historically obsessed with discussions of love.

> Indeed, people have done nothing but speak about love in psychoanalytic discourse. How can one help but sense that, with respect to everything that can be articulated now that scientific discourse has been discovered, it is purely and simply a waste of time? What analytic discourse contributes ... is that to speak of love is in itself a jouissance.[23]

Here Lacan locates a particular pleasure in the process of 'speaking' of love, which the psychoanalysts have historically tended to do plenty of. While generations of psychoanalysts have treated the topic of love with their patients, for Lacan this discourse has been a 'waste of time' when it has been forgotten that the lesson of psychoanalysis ought to be that there is a particular kind of jouissance – an important formative pleasure – created by the very act of speaking about love. Speaking should not be taken too literally here, since for Lacan to speak is also to enunciate, even to be *spoken through* as if the subject is the effect of a discourse rather than simply its *cause*. In other words, to understand love we also need to understand how we speak about it, and how it speaks through us.

In *A Voice and Nothing More*, one of the most important books on psychoanalysis written in recent years, Mladen Dolar argues that there is a 'minimum of ventriloquism' in every speaking subject. To paraphrase Dolar's argument, it is not so much that the subject is an active agent *who speaks*, but rather the subject *is spoken* by and through a kind of ventriloquist discourse whereby the subject appears to be the speaker whose voice comes from within when they are in fact the agent of another voice.[24] This idea of a pleasure yielded from being *spoken through* will help us understand how virtual reality participates in a vital political transformation of the love and sex industries. If love and desire speak through us, they do so in a particular way when it comes to the digital technologies of today.

While it may seem a rather more abstract point, both the idea of a pleasure yielded from the process of being involved in a representation of love described by Lacan and the kind of ventriloquism that Dolar describes will help us to understand how the subject is able to experience the immersion of a VR simulation when the subject of that simulation is desire (dating simulators, AI chatbots, VR pornography) as if it is related to the desire of the user themselves. Lacan stresses that desire is always 'the desire of the other', which means not so much that we desire another person or thing (as general discourse assumes) because of our specific and unique connection to that object in question but that we get our desire because it is first a desire that exists elsewhere.

It's an argument interestingly anticipated by Simone de Beauvoir when she argues that 'desire for love becomes passionate love'. Speaking of sex acts, de Beauvoir writes that a more abstract desire *for* love can be replaced by passionate love for the individual with whom sex has taken place. In this sense an existing desire – one existing elsewhere not precisely in another individual but in a wider milieu constructed by its surrounding politics – becomes the desire of the subject. As Roland Barthes once wrote, 'no love is original'.²⁵ Desire, then, has an almost imitative (memetic) quality, much as we could say of knowledge. In other words, like with knowledge, no one person can know something (or any incantation of it) in isolation, because in that case it would be mad, inarticulate and even unsayable. For knowledge to emerge, it must be understood by more than one person. Like knowledge, for desire to emerge there must be a form of agreement in place – a cultural, social and political structure – which has made that particular desire possible in order for the subject to, as it were, *opt in* to experiencing it.

These arguments show that psychoanalysis should not be seen as a search to uncover or understand common desires rooted in humanism (we all feel X or we all have Y in common), a criticism that has been lobbied at psychoanalysis, but as a theoretical perspective that shows us that desires become common by means of a ventriloquism of desire, a ventriloquism that is now governed by the viral possibility of the digital: the digital as an infinitely reproducible media of viral desire-experiences which not only respond to but construct the desire of its users. It also shows psychoanalysis as the ally of anti-capitalism, since it is the means of revealing how our desires are constructed and reconstructed by patterns of digital development and its associated economies, a fact that capitalism and its world of desirable commodities is eager to hide.

We should use psychoanalysis to understand the experience of *Summer Lesson* then, or to make sense of an hour or two spent on a VR porn site like Czech virtual reality, but we should not do so by assuming that the experiences are about tapping into some kind of unconscious and, importantly, already-existing desire present in the user. Instead, we should see these experiences as constructive of an imitative desire, offering the user the pleasure (*jouissance*) of a new scene of desire, a pleasure that maybe a temporary departure from their everyday life or a more significant influence on that subject's desire in a longer-term way. Chloe Woida (2009) makes the insight that mainstream online porn seems 'as if made for someone else', noticing this important psychoanalytic structure of how we experience these digital invites to enjoy.²⁶ This would intensify the problem of homosociality in the industry identified by Paasonen just above, since the 'desire of the other' which the subject is invited to experience by ventriloquism is coded as a particularly masculine one.

This approach offers an alternative to the bizarre for and against argument that has emerged in academic responses to sexbots. One of the most high-profile interlocutors in the debate is the Campaign Against Sex Robots, started by Kathleen Richardson, professor of Ethics at De Montfort University. Her paper on the topic argues for a deep connection between sexbots and prostitution, arguing that 'the development of sex robots will further reinforce relations of power that do not recognise both parties as human subject'. Among other things, Richardson argues, sexbots will intensify the objectification of women and children, encourage a culture of prostitution and even intensify the prominence of violence in sexual relationships.[27]

On the other side of the debate is Kate Devlin of Goldsmiths University, who argues on the contrary that sexbots might have liberating or even emancipatory potential for the future of sex. Located more in the history of sex toys and practices than in the development of digital relationships that is our topic in this book, Devlin discusses figures like Betty Dodson, a 'pro-sex feminist' whose decades of work have contributed to the diversification of sexuality in line with progressive politics, and she sees sexbots as a form which could at least potentially contribute to this progressive movement.[28] Brothels like Unicron, mentioned above, share something with this approach.

Cutting against these two dominant positions in debates surrounding the ethics of sexbots, Isabel Millar offers another much-needed psychoanalytic perspective. In her 2019 article 'Sex-Bots: Are You Thinking What I'm Thinking?', she writes:

> The common-sense approach to sex that seems to be apparent in these two above arguments [those of both Richardson and Devlin] for and against sex-bots, is that there is something inherently meaningful and even *natural* about the way that we have sex. And by natural, I mean 'instinctual' or biologically programmed. The Lacanian psychoanalytic approach to sex is entirely more pessimistic and indeed suspicious of such easy explanations.[29]

There is a significant connection between sex and videogames here. The two dominant positions in academic and general debate about the relationship between games and violence are repeated here, with the same problematic assumption about the apparently 'natural' underlying impulses which games respond to. The now almost defunct argument that videogames risk the user acting on underlying violent drives and the alternative argument that games offer an outlet for otherwise dangerous drives that might otherwise be unleashed upon society both suffer from this misconception. This link between assumptions made about games and those made about sexbots might be more than incidental. The presiding narrative of Silicon Valley developments like those of the start city discussed in Chapter 2 is that these technologies give the user what they want. This this dominant line offers the company a kind of diminished responsibility (e.g. in OkCupid's response to the radical biases of their algorithms) and without it the politics of the technology design would come under far greater scrutiny. In fact, addressing this could cut against the whole neoliberal logic of the 'Californian Ideology' that emerged from the dotcom neoliberalism of the 1990s and still holds oddly significant sway in discourses around technology today.[30]

Instead of these two blind perspectives which risk seeing desire as natural, Millar argues via Lacan that 'sex is completely artificial' and it is only by recognising this that we can begin to think about the politics and ethics of developments in sex and love.

> The reason why sexbots are so problematic and yet so fascinating is that they expose precisely the artificial character of the sexual relation. In Lacanian terms the unbearable *real* of sexuality. The fact that an artificial doll may act as the representation of a sexual fantasy presents us with the true horror of subjectivity: the fantasy is the only thing that really sustains any of our relationships at all. This nugget of *artificial wisdom* is Lacan's lasting legacy to us as we move forward into the realms of AI sex and love.[31]

What sexbots also show us is that the fantasies that sustain the existence of our desire are not our own. It would be easy enough to disagree with this position simply by retreating to a narrative of humanist or instinctive desire, but even if we consider elements of desire and impulse to have their roots in natural or biological condition it by no means follows that we are experiencing desire in an entirely natural way in any of our contexts today.

DIVIDED DESIRES

It should be clear enough that desire is at least mediated, edited and transformed – if not entirely constructed – by the digital processes discussed so far. Beyond the politicisation of desire that such a realisation makes essential, psychoanalysis here offers us the chance to make visible the construction of our desires, and perhaps even to change them. If desire is a kind of memetic fantasy, as per Lacan and Millar, then we are less desiring-machines (the famous term of Deleuze and Guattari from their critique of psychoanalysis in *Anti-Oedipus*) than we are machines through which desire can move. Memetic desires, moving like viral memes through their machinic subjects, affect and infect subjects with their politics and biases in full force. What is left after making this realisation is to take active part in the construction of these memetic desires with a political awareness in the future.

Memetic desire such as this might provide opportunities for empathy and identification with others by allowing for a shared experience of libidinal energy. In an influential argument about videogames Alexander Galloway argues that the unique quality of games (compared to films, for example) may be that 'where film uses the subjective shot to represent a problem with identification, games use the subjective shot to *create* identification'.[32] In other words, games make us identify with certain positions, removing some of the critical distance that is implied in other forms of media. This can make these experiences radically influential, both when they are serving hegemonic or dominant ideologies and when they deviate from these norms to produce new experiences of subjectivity, as we have seen in many of the examples discussed above.

On the other hand, our experiences online and in digital life can be very fragmented. Often, our screens demand that we move from one thought, feeling or desire to another with high frequency. We switch rapidly from meme to meme, from game to game, from one simulated burst of libidinal energy to the next. The result of this might be to produce an experience that is divided rather than consistent. In 'Postscript on the Societies of Control' Deleuze introduced his concept of the 'dividual', which has in recent years become a staple term for discussions of the subject in his/her digital environment. The 'dividual' is the idea of the individual as a collection of data but also as a being who can only follow certain predetermined paths, just as computational machines can. As opposed to an individual, imagined as a consistent identity who stays the same day to day or who changes in a continuous way over time, the dividual is an experience of ourselves in which we are encouraged to restart and reorganise ourselves regularly in a ruptured or fractured way. The possible paths that we can follow are refreshed in each moment.

The different internments or spaces of enclosure through which the individual passes are independent variables: each time one is supposed to start from zero.[33]

For Deleuze, when we pass into these experiences we are 'reset', experiencing the moment in isolation or before moving on to the next. As such, we become subjects characterised by division and flux. This situation might create temporary moments of identification – even of empathy – but they may prevent rather than encourage the longer-term development of solidarity and fidelity. We are not only divided from others but from and within ourselves. In this theoretical context it is easy to see how virtual reality events have been used so successfully to attract consumers and investors in venture capitalism and big business, such as companies like Aures London, or even to encourage charitable donations to particular philanthropic causes. By momentarily capturing attention and libidinal energy, such experiences foster temporary identification with the ideology in question. What might be less likely is the creation of long-term support and solidarity to develop through such acts of digital identification. Nevertheless, we should try.

PITCH – PLAYPOL, THE POLITICAL DATING SIMULATOR

Pitching a new conceptual dating simulator is about as tricky as it gets were we to be on *Dragon's Den*. As discussed above, there are nostalgia dating simulators, ghost dating simulators, animal dating simulators and pretty much any other variation on the genre imaginable. When it comes to the form and gameplay of such games, however, this superficially diverse range of software has strong common features. They replicate, in general, the kind of playable pick-up artistry of the dating examples discussed above. They all thereby rehearse variations of what we might see as the dominant or normative power structures of imaginary sexual life, even when they might occasionally pitch themselves as having parodic or even 'woke' content.

If we look beyond the content or stories told in these games to the level of algorithm and interface and the structures of playability that have increasingly gamified dating over the last decade, there are a number of ubiquitous trends across the range of varied simulators. They all either emulate a monocular gaze of the first-person shooting genre or construct a fictional playable avatar whose story the user controls for the entire experience. In the first of these instances, prominent in both VR and mobile simulators, the user imagines themselves looking into the screen rather than being seen appearing on it, inviting the user to see themselves (and thereby their own desires) thrown into the scenes of desire that are happening on screen. In the second category, where the user plays as an on-screen avatar, the user is perhaps confronted more directly with the memetic kind of desire discussed above, temporarily experiencing the desire of the imagined subject who desires the imagined object.

To think seriously about how a dating simulator could function for progressive politics we first need to think about this core pleasure of the experience of having a stand-in or imagined object of desire replace, usually temporarily, the desires imagined to be one's own. The avatar-driven desire provides a clue to understanding these processes. We could easily understand the subject's readiness to replace one object of desire for another, simulated or otherwise, though the Lacanian idea of the *object a*, the endlessly elusive and inarticulable object of desire which constructs the subject but must always remain out of reach. Everything the subject desires – from the Pokémon to the sexbot – can be seen as a stand-in for this illusory object.

But the picture is more complicated. Such readings of Lacan's model for desire risk implying a fixity in the desiring subject itself (the objects may change, even hourly, but the subject appears relatively consistent in the formulation). The kinds of dating simulator, from videogames to AI chatbots and sex robots, that have been discussed in this chapter clearly show something more complicated. They show that the libidinal pleasure of these experiences it is not simply in the object but in the pleasure yielded from taking up another subject position in relation to the object. In other words, it's not so much that we want the sexbot or the VR snog but that we can derive pleasure from taking up the position of a subject who wants these objects. This conception fits with Freud's initial conception of the 'drive' when he writes in *Three Essays on the Theory of Sexuality* that:

> The simplest and likeliest assumption as to the nature of [drives] would seem be that in itself [a drive] is without quality, and, so far as mental life is concerned, is only to be regarded as a measure of the demand made upon the mind for work.[34]

The drive is to be thought of as an invitation to the subject to put its psychological life to work desiring. Freud always insisted that we should think of drives as 'partial' and this was something picked up by Lacan. In Lacan's view drives, like the oral and the anal drive, are partial in that they do not make up one tendency in people's functioning and there is no completely satisfying object connected to them. Rather, they confront the individual with 'turbulent movements' in his own organism and give rise to a fragmented experience of the body.

Furthermore, the digital object of desire cannot be seen simply as replicant or as inadequate stand-in for another object, even if it does operate psychically as a replacement for the ultimate object of desire. Quoting Adorno, Illouz notes that the 'commodified exercise of imagination is a central dimension of a modern bourgeois consumer society' and that 'this institutionalisation has transformed the very nature of desire in general, and romantic desire in particular'. For Illouz, economic and cultural institutions:

> have clearly codified the cultural fantasies through which love as a story, as an event, and as an emotion is imagined, and it has made imaginary longing its perpetual condition. As an emotion and a cultural cognition, love increasingly contains imaginary objects of longing: that is, objects deployed by and in imagination.[35]

Illouz approaches her argument through this concept of 'imagination'. She notes two contradictory functions of the word. In one sense (an idea pushed by Jean-Paul Sartre) imagination is a pale replica of what can be perceived by the real senses (if you close your eyes and imagine something, it will be an inadequate or incomplete version of what your senses are capable of perceiving). In another sense, imagination has been seen primarily as something which 'takes hold of the mind far more intensely than ordinary sense perception', something which embodies our 'capacity to invent something that was not there before, to magnify and intensify our lived experience'. Taking in both these contradictory meanings, imagination is both an inadequate copy of what we might call reality and an idealised dream of what we imagine we want reality to be. It responds to reality but also takes part in the future construction of it. For Illouz, 'perhaps nowhere more clearly than in love can we observe the constitutive role of imagination: that is, its capacity to substitute for a real object and to create it'.[36]

In the kind of dating simulations discussed in the chapter above, this metaphorical replacement of the real object with an imaginary one (perhaps a feature of every relationship) is embodied in real and clear terms. One final point to take from Illouz is that 'imagination here is both private/emotional and social/economic'.[37] Likewise, in the history of dating simulators – from *Girl's Garden* to *Summer Lesson* – and like all the examples of libidinal technology discussed in this book, the pleasure of the user is both a political and economic construction of power and an experience *felt* as private, emotional and responsive. It's also an active construction of the imagination-to-come, a construction that can be reconstructed. In properly psychoanalytic terms we could formulate this the following way: the relationship of the *object a* to the daily lived experience of partial drives is one determined by the imagination and its social and political constructions.

The task of the political dating simulator would be to confront the subject not only (or perhaps even) with recognising the memetic nature of their desire and its political and economic roots but to explore the possibility of making libidinal transformations to their own subjectivity for political reasons. It could – at its most effective – reorganise the relationship between the subject and its partial drives in the service of progressive politics. To do so, it would need to be working on the level of form, gameplay and experience, offering the player a new form of play entirely. Some existing dating simulators – and a great many more videogames – communicate progressive values in their narrative and content – but if they are to go further, we must reimagine how a simulation could function psychically on its player.

Liberal themes in videogames, from *Detroit*'s 4,000-page script assessing the dangers of AI to *Assassin's Creed*'s narratives in support of ethnic minorities and oppressed groups to the queer dating simulators discussed above are significant. While many cultural critics point out that games tend to support militarist or conservative values, on the vocal right-wing of the gamer community, the claim is regularly made that games extol left-wing ideology by featuring marginalised characters: in short, any character that is not white, straight and male. But this liberal influence has failed to turn videogames into a force for progressive politics.

In short, progressive content is not enough. Shooting games like *Wolfenstein* might be about killing Nazis, but it gave birth to the first-person shooter genre, in which players often spray bullets in the service of

American foreign policy. City simulators like *Civilization* and *Tropico* might allow identification as a socialist state or egalitarian democracy, but they require adherence to the principles of Western capitalist empire-building to succeed on gameplay level. Videogames communicate ideology at the level of form, and laying a progressive storyline over the top does not necessarily prevent a game from serving the hegemonic kinds of ideas discussed in this chapter. With dating sites, the users are the content, so a progressive dating site would need not only to feature progressive individuals (as Red Yenta may do) but it would need to be structured progressively at the level of form: its algorithms.

Welsh Marxist Raymond Williams showed – years after the fact – that the form of the nineteenth-century novel still supported conservative values, despite containing left-wing content. For Williams, novels such as those of Charles Dickens might sympathise with workers or with revolution – even lionise them – but they could never instigate serious social change, because their political power was limited by the fact that the books were still a commodity to be read after dinner and for pleasure: a kind of storytelling that is easily consumed and doesn't make the reader confront real politics. The same would apply to many videogames today. Eventually readers and writers in the early twentieth century grasped this, and as literature moved towards modernism, writers realised that the form needed changing – rather than just the content – if literary culture was to become progressive and politically forward-thinking. Modernist novels refused to be comforting, introducing stream-of-consciousness techniques and unreliable narrators that unsettled the reader and made them confront the politics of their own relationship to the text. This kind of shift still needs to happen with videogames, and it needs to happen with dating technologies.

Content is not inconsequential. Some games succeed in using content to comment critically on their form. One example is *Sweatshop*, a management simulator that confronts labour conditions. Another is *Cookie Clicker*, an automation simulator that ends with an accelerationist collapse of capitalism. Both games subvert the form of management games and show it up. *Autosave: Redoubt* likewise subverts *Counter-Strike* to highlight monocular politics of the first-person shooter, questioning how games frame the world from the predatory gaze of the phallus/gun. The next step is to go beyond critiquing existing forms and produce new ones. It's perhaps impossible to predict what a formally progressive videogame might look like, just as we could not have second-guessed the radical modernist novel before the works of James Joyce and Virginia Woolf. What we can see already, however, is a number of emergent games which experiment with form in potentially game-changing ways. Perhaps we will soon see videogaming's Ulysses. For now, we'll attempt something similar for a dating simulator.

The existing game that provides the best blueprint for a progressive form of dating simulator is David O'Reilly's 2015 ecological exploration game *Everything*. It has nothing do with dating, but has a unique and experimental gameplay style in which the player starts as one species exploring the world but is encouraged to endlessly switch to the perspectives of other objects and species until the boundary between 'subject' and 'other', so solid in the majority of games, collapses. If *Autosave: Redoubt* points to a problem of perspective in games, where events are viewed only through the eyes of one agent (the player-character), *Everything* creates an alternative one.

This idea of switching between subject positions in simulated dating, sexual and relationship experiences could form the core concept for a new dating simulator which would work to connect desire, politics and play in a new way. In all of the examples of dating simulators discussed in this chapter, the simple change of switching from the playable character to the object character being played would radicalise the experience immensely. In *Summer Lesson*, the playable English teacher flirts with a young Japanese girl alone in their country mansion miles from a city, alone in an erotically charged paradise. A simple switch of perspective so that the user became the young student would radically transform the experience. Suddenly the scene would be completely recast, the paradise island suddenly visible as a vulnerable isolated space cut off from recourse to help, the country retreat now a workplace, the flirtatious atmosphere a scene of danger. In *Super Seducer*, the user coaxes women to a hotel room by using their financial clout and a simple switch of playable perspective would make the transactional nature of the experience visible. In some VR kissing simulators, the user has to move their head forward at the right time to seize the chance to get a kiss, framing physical movement at a moment of vulnerability in the other as a normal part of romance. A switch of perspective mid-lean would confront the lack of active consent in such an act. With this simple technique, the homosociality of the monocular framing of these experiences could be undone.

If it were to stop here, the simulator could be a kind of #MeToo The Game to combat *The Game* itself, making visible the amount of sexual politics and scope of power relations that embed themselves into so many of our daily romantic experiences – that dominate what Illouz would call the romantic imagination – and which are taken for granted as 'natural' features of desire. More could be done. Rather than playing two sides of the same romantic experience, players could be invited to play a series of seemingly disconnected 'scenes' of desire. The range would show the user the memetic nature of desire, divorcing the experience from the illusion that desire is consistent, unified or apolitical. The player could be invited to 'rank' these scenes of desire in terms of the most provocative in a kind of mock-up of PornHub's voting algorithms, which would make visible how the sexual industries have operated with a data-driven filter bubble (more in the next chapter) to naturalise the desire of the individual user and produce the image of its ubiquity while hiding the existence of diverse desires from the stream.

In a further feature, the game would ask the user to play two sides of the same exchange but without making them aware of it at the time of playing. By inviting them to play the same form (structure, algorithm) with different content (images and narrative) the simulator would take the user through a romantic experience from both sides without their knowledge. In other words, the same structural scenarios repeat themselves but disguised under different narratives and with different playable avatars, seen this time from the other side. This would make the politics of algorithms (and of romance) visible and at the same time allow the user to absorb themselves – displace their desires – into the same structural romantic situation from two sides without them being prejudiced in advance by the desires experienced in the first play-through (whichever side they played as first). At the end of the sequence, the game will collect input about the user's experience of the situation before revealing which scenes are paired, forcing the viewer to confront what Illouz identifies about the romantic imagination, that it is both private/emotional and social/economic, driven by economic factors but implicated in personal impulses. A second play-through would then test whether the experience of the game had influenced the pleasures of the player. The software could be playfully connected to the kind of wearable discussed in Chapter 2 to record responses via heart-rate to contextualise bodily affective reactions to romantic images. This would operate almost like the 'smart condom' developed by iCon, but where that can function to collect sexual data to compare on male-oriented blogs full of sexual boasting, this would show the subject its own libidinal reaction to situations charged with political and cultural stimulus.

Ultimately, creating a political dating simulator would be designed to confront the line between simulations and representations and their associated desires, showing that the cultural imaginary is actively constructive of how we desire and can be changed along with the economic structures which support it. A long history of digitising the world of romance has created a cultural revolution in how we relate to our lovers.

Building a progressive dating simulator like this would hardly change the world, but it would at least cut against this 30-year history and reverse the trends in gaming, mobile apps, VR and dating platforms that are dominant today. It might have a small role to play in the desirevolution that is in full swing.

4

The Match: Metaphor vs Metonymy

'My phone became a tiny casino, an oracle, a source of habit-forming terror and elation'

– Roisin Kiberd, *The Disconnect* (2021)

Logging in to your dating app on a Friday lunchtime, you wistfully scroll and tap your way through the pages, waiting for that screen of enamoration to appear with a pulse of desire and give your weekend ahead that invaluable aura of hope and possibility that you feel it deserves. There's an element of chance, and you know that you may not be in luck today, but there is also a great deal of hope. What you probably don't realise is that your phone knows full well just who it is that you want to see. It has several screens of enamoration to show you, which it has ranked and curated in terms of the likelihood that you will feel that powerful pulse of desire when the screen finally appears in front of your eyes. It has the power to turn your weekend around. It has identified the most likely combinations of data and earmarked them with a special tag to separate them out from the infinite screens of mundanity (the digital wheat from the arithmetical chaff) which you have been scrolling through since lunch, for it's now half-way through your afternoon and your mailbox is full. The data knows what you want, but today your phone has withheld these screens from you. It has done so because you took a very busy route to work this morning and chose not to wait for a less congested train, and because you scrolled past a message of support for Julian Assange that one of your brother's friends posted in their social media feed as you were coming round before your morning coffee. You have carried out a minor act of selfishness and associated with the wrong person, and your government has docked your points score just below the premium level you need to be allowed access to the love of your life (or at least the love of your weekend).

DESIRE DYSTOPIAS

You are not living in the dystopian future, nor have you been cast in an episode of *Black Mirror*. You are just one of 85 million users of the dating site Baihe (Tantan, the most popular dating site in China, has 270 million users). Many dating sites now offer paid-up premium versions (Tinder Gold, Hinge Preferred, etc.) and some offer a kind of loot boxing strategy (you can buy 'beans' on the app Coffee Meets Bagel which allow you to bypass some of the time restrictions on the dating game). Baihe takes things a step further by using the Zhima Credit score, a complicated credit-checking and performance ranking system developed by Ant Financial Services, a company closely affiliated with the Alibaba Group. Here, a centralised ranking system determines who gets to love and whose desires will be considered for fulfilment. The program was implemented in 2015 and was one of the first credit ranking systems introduced in China. The system works initially like a standard credit check and was modelled on the American FICO score that is used by most banks and credit grantors in the US. The Zhima Credit system takes the credit check to new levels though, using not only the user's credit history and their record of returning successful payments and re-paying debts but a range of complex personal information which can be recorded and ranked to award the user points or issue penalties against them.

One category the score considers is 'behaviour and preferences' which includes making a judgement on the user's tastes and movement and reading their browser history to see 'websites visited' and monitoring 'product types' the user purchases. An even more concerning category is 'interpersonal relationships' which involves taking a record of the user's friends and connections and then making a judgement on 'the online characteristics of a user's friends and the interactions between the user and his/her friends'.[1] In other words, even association with low-scoring individuals can decrease the user's score, essentially privileging those who socialise only within higher economic circles. This not only takes surveillance of citizens to levels even Cambridge Analytica hadn't dreamed of but it turns the trend of cyberbalkanisation – the splitting of the apparently equal and open internet into closed groups (loosely referred to as filter bubbles) based on class, race and other identity categories – into a formal process and punishes the user for breaking across these lines.

The Zhima Credit experience becomes another strange form of gamification where to succeed the user must climb a ladder of social class and government conformity while ensuring their friends and acquaintances do the same. Users are scored between 350 and 950, with most users maintaining a score of over 550. People with over 650 points are usually white-collar workers, while those with more than 700 are commonly described online as future millionaires. Users above the 750-point mark are heralded those destined to become a billionaire within a decade. Many Chinese citizens report actively working on their score and taking actions to improve it both for fun and for the benefits it might bring, making it an embodiment of gamification. Among the rewards are faster visa travel paperwork applications, medical treatments on credit, speedier airport security checks, access to loans and car and home deposit discounts and waivers and access to credit cards, as well as the ability to connect with buyers and sellers on marketplace platforms like Xianyu and with potential lovers on Baihe.

This stratification of the internet into identity groups is the very logic of how social media sites introduce new people to one another and of the algorithms used by the 45 dating sites and apps in the IAC Match Group. It was almost a natural development then when the dating site Baihe started using the Zhima Credit score to help rank and match those looking for love. When put in terms of dating this can be seen as stratification of desire along class lines, organising society into conformism no longer by traditional marital structures, as it was for philosopher Claude Levi-Strauss in *The Elemental Structures of Kinship*, but by preventing cross-class and diverse identity solidarity by ensuring a carefully splintered internet of desire – a libidinal splinternet – in which subjects and their objects are organised into manageable herds secured by invisible but powerful fences. One advantage for platform capitalism is that these herds can be monetised and targeted by information campaigns. Another is that keeping people apart in their bubbles prevents organised widespread solidarity, and potentially even revolution.

It is perhaps not new that love has always been at the heart of the capitalist organisation of people and things. Connecting the historic connection of love and capitalism to online dating services, Todd McGowan writes:

The commodity that the dating service sells is much more valuable than those sold by the grocery store because it carries with it the illusion of a complete satisfaction. No one believes that eating a particular kind of ice cream will provide such a lasting satisfaction that I will never desire ice cream again, but many in capitalist society believe that finding one's soul mate will permanently solve the problem of desire for a love object. This difference bespeaks the pivotal role love has within capitalism. It is not just one commodity among many but the central commodity. One might say that all other commodities are modelled on the love object rather than vice versa.

McGowan is right that love is at the heart of a capitalist organisation of society, and that the 'dating service tries to mask the unexpectedness of love by making it thoroughly predictable [and] transforms love from a disruption into a stable structure for one's life'.[2] For McGowan, while 'love' might be disruptive and potentially subversive, 'romance' refers to its capitalist organisation. What is important to add here is that the development of these technologies of love and romance shows they are not just interested in abstractly organising the disorganised or fragmented nature of love and desire into something that can make manageable sense. Rather, they are specifically interested in organising the desire economy into particular class-oriented patterns to set the blueprint for a capitalism to come that is clearly demarcated and stratified along economic lines. To put it simply: it's not just that love is mad and technology struggles to make sense of it, it is that the organisation of love is designed to get control over it to serve particular economic interests.

THE LIBIDINAL SPLINTERNET

Another of Zhima Credit's dystopian but fascinating categories is the idea of measuring a user's 'honesty'. This relates to the way the internet has gradually been transformed in the 30 years since its inception from a space of radical experimentation to one of consistency and regulation. The score verifies the 'honesty' of the user's online information, ensuring that their online profiles stick to a consistent and truthful account of the individual it claims to represent.

This potentially formalises and secures the more gradual changing shape of identity on social and online media since the internet became a common feature of most households in the mid-1990s. In 1996, Sherry Turkle published her influential article 'Who Am We', arguing for the liberating and democratic virtues of the newly populated internet.[3] It was

before the birthday of Microsoft's first browser – the original Internet
Explorer – and this was a time of hope and optimism for a new form
of social life online: a utopian dream in which the power of traditional
institutions and media outlets would subside to make way for a more
democratic and diverse marketplace of ideas in which individuals were
empowered to create, share and discuss in ways never before imagined.
Even in 2008, when Barack Obama was elected via a grassroots movement which mobilised the internet in unprecedented ways, this dream
was – at least to a certain extent – alive. Chapter 1 considered the data-
driven elections that have changed the shape of this grassroots potential
internet in political terms, but there have been equally dramatic changes
at the level of self-presentation and in particular for the process of building relationships and friendships online.

At the heart of Turkle's argument was the idea that this diverse and
multiplying online space, composed of social media pages, multiplayer
videogame dungeons, personal blogs, websites and collaborative projects,
would be a space for experimentation with identity and escape from the
normativity of identity in offline life. In some ways her approach might
be intuitively comparable with Donna Harraway's feminist argument in
her famous 'Cyborg Manifesto' in which our digitisation represents an
opportunity for a new and less patriarchal and heteronormative politics.[4] Turkle's article discusses a range of individuals who were taking
advantage of these opportunities for experimentation in the mid-1990s,
including members of the LGBTQ+ community whose offline circum-
stances were less than supportive of their gender identity, potentially
polyamorous or other non-monogamous users whose sexuality had little
contemporary support and various users whose diverse desires for role-
play and other non-normative practices might have fallen outside of
common contemporary behaviour. The early internet, then, seemed a
place for more flexible identities, and in fact the Usenet newsgroup alt.
polyamory created in 1992 is retained in the Oxford English Dictionary
as the first appearance of the word, showing a close connection between
early connective and social online communities and innovations in relationship building and sexual experimentation.

By 2010, the picture was very different. In that year Mark Zuckerberg famously gave a speech in which he proclaimed 'you have one
identity' and argued that it represented a 'lack of integrity' to have one
identity for work colleagues and another for private life. Jose van Dijck
has shown how central this movement in the fabric of social media has

been.[5] Facebook would begin its project of digitising the lives of its users into manageable strands of data whose apparently consistent identities would allow for personalised newsfeed curation and directed advertising. The exposure to diverse information flows imagined by Turkle and her fellow optimistic early surfers would be replaced by a handover of decision-making to the platforms themselves: the Youtubes, Facebooks and Twitters whose algorithms would decide who sees what, and who meets who, on the basis of each user having a single identity represented by a profile which can be compared against the 'truth' or 'authenticity' of that user.

In 2019 Facebook – penetrating more than 28 per cent of the globe, more than any traditional media outlet – announced its news authenticator feature through which 'deeply-reported and well-sourced' news will be specifically marked out on the platform. Separately, it announced an update which will increase levels of censorship on the platform. Apart from the questionable decision to include sites like the far-Right Breitbart – which is ultimately little more than a privately funded collaborative blog – as a legitimate news source, this moment marks a shift of power away from traditional media. However, this power has moved not into the hands of the people, as early proponents of the internet had hoped, but into the hands of the platform capitalists who now decide (as the newspapers and TV channels once did) what is legitimate information and what is not. While this development made the news, it's precisely this process of determining legitimacy that Facebook had been carrying out on its users' identities for a decade since. When a company like Zhima Credit uses this 'one identity or a lack of integrity' model as a basis for success in its ranking system, this trend of influencing the online subject into a single predictable data strand who can be matched with products and services using detailed psychometric profiling becomes formalised. This same data structure is also the basis on which users can be matched with potential lovers.

What we are looking at here is a libidinal version of what has variously been called filter bubbles, the splinternet and cyberbalkanisation, the process by which digital space is organised and curated into distinct blocks in which specific demographics of individuals circulate and cycle around a web of hyperlinks and connections but rarely – if ever – step outside the borders of that digital space. Today we are living in thousands of online micro-economies of desire in which users loop around in cycles of pleasure which are directed to generate corporate profit while also preventing the development of a mass culture of solidarity and resistance. The proposal for a new dating site at the end of this chapter seeks to reverse this structural aspect of online life.

SMV: SEXUAL MARKET VALUE

This may connect rather worryingly with what in some circles can be called SMV, or sexual market value. In her 2020 book *Going Dark*, Julia Ebner explores and infiltrates a range of far-Right and extremist groups, including the online male supremacist group Men Going Their Own Way, abbreviated as MGTOW and associated groups of women known as Red Pill Women or Trad Wives, groups which have over 30,000 members. Most women who join these groups, Ebner explores, do so in 'the search for love'. The groups perceive gender roles as the result of a 'sexual economics' in which the heterosexual community should be seen as a 'market place where men buy and women sell'. The SMV is a score that men can assign to women to designate their value in this market place.[6] The tendency to use scores, ranks and numerical values in these communities goes further still, with each woman being given an 'N-count', referring to the number of sexual experiences she has had (considered a key piece of data in determining her value) and with each existing relationship given an RMT (relationship market value). The case is a perfect example of a micro-economy of desire that characterises the libidinal splinternet.

These communities are close to those of the pick-up artists discussed above and there appears to be a return to such ideas not only in the realm of gaming and play, as discussed in Chapter 2, but in other online cultures such as these, and perhaps even at the heart of online dating itself. In 2017 DeAnna Lorraine, a 'red pilled' dating coach, wrote the vanity press book *Making Love Great Again: The New Road to Reviving Romance, Navigating Dating, and Winning At Relationships*, which is essentially a guide to sex and love from the perspective of an alt-Right convert. The book's title seems to be a nod to the Trump.dating site, though the book itself doesn't mention that project. Nevertheless, though it criticises the MGTOW movement, it is a combination of dating tips with tirades against contemporary feminism, endorsements of Trump's presidency and a bibliography full of far-Right and traditionally right-wing writers and publications from Breitbart to the *Daily Mail*.[7] The readership for such a book might be relatively small, but it connects to a much larger online community that combines a gamified approach to dating with a political agenda. If the strategy of Zhima Credit has its own politics, one of conformism and dedication to economic stratification, there are other significant pockets of the dating world which politicise ranking and gamifiying the dating world for more extremist agendas, on the Right at least. What both the far-Right subcultural activists and the neoliberal tech conglomerates share in their triangulation of love, data and value is the enshrining of money or value as both a barrier between the subject's desire and its fulfilment and (paradoxically) as the mechanism through which desires can be fulfilled. If the subject had a higher SMV, or a better Zhima Credit score, then desire could be fulfilled, or so the assumption goes. The whole logic of the 'incel' (involuntary celibate) movement is to place an imagined barrier between the subject and its desire.

This process of connecting desire with money was identified by another of the most important figures critiquing contemporary capitalism as its early characteristics emerged at the beginning of the twentieth century, Georg Simmel. Simmel was a sociologist and a philosopher of capitalism, of the city and of psychology. His most famous essay was his 1903 text 'The Metropolis and Mental Life', which argued for the city as a force for reorganising psychological processes over a hundred years before the technologies deployed by the smart cities of today. Simmel was an influential figure in the development of Marxism. He was an associate of Max Weber, a significant influence on the Frankfurt School and György Lukács claimed to have come to Marx's work from reading and studying under Simmel. It was in his huge 1900 book *The Philosophy of Money* – which has been seen as an extension of Marx's *Capital* – in which Simmel developed his theory of money and desire.

For Simmel, desire as we experience it in capitalist society is connected to commodification. Yet, it is not simply that capitalism organises or harnesses desire for its own agenda but that our libidinal relationship to objects and other subjects is fundamentally constructed by capitalist society. To grasp this, Simmel begins with a conception of love from Plato:

> This dual significance of desire – that it can arise only at a distance from objects, a distance that it attempts to overcome, and yet that it presupposes a closeness between the objects and ourselves in order that the distance should be experienced at all – has been beautifully expressed by Plato in the statement that love is an intermediate state between possession and deprivation.

Anticipating the psychoanalysis that was in its embryonic form just hundreds of miles away, Simmel sees desire as something that entails a connection that is also a separation, a separation that is needed in order to feel the possibility of the connection. To love is to be caught between this 'possession and deprivation' that Plato describes, experiencing a draw towards an object of desire which is only made possible by the feeling of a gap between desiring subject and desired object. In other words, we don't simply want the object – be it a lover, a Pokémon or a car – but we want it in and because of our separation from it. The concept could be explained by the love of celebrities, for example, where the seemingly insurmountable distance between the desiring subject and his/her object is constitutive of the desire itself. What Simmel adds to Plato is the entry of value and money into this mechanism of desire:

> The subjective events of impulse and enjoyment become objectified in value; that is to say, there develop from the objective conditions obstacles, deprivations, demands for some kind of 'price' through which the cause or content of impulse and enjoyment is first separated from us and becomes, by this very act, an object and a value.[8]

At some imprecise historical point, price or value enter into this complicated desire equation. Money – or an equivalent value – begins to operate as the mechanism which appears to connect subjects with their objects both as the means of accessing the desired object and as the prohibitive barrier between the subject and its acquisition. The Zhima Credit score on Baihe might be an almost comical embodiment of this process, but it would explain our relationship to objects of desire in a much wider sense as well. To continue with the example of celebrities: the insurmountable gap between us and the object can only be surmounted by a dedication to getting rich, which would suddenly remove the barrier between us and the fulfilment of a desire, or so we imagine (we don't literally save money so that we can marry George Clooney, but we buy a Nespresso machine as if we might). In this way the object of desire itself becomes an object in a new way, an object with a particular relation to the economies of value. Simmel continues:

> The fact that I want to enjoy, or do enjoy, something is indeed subjective in so far as there is no awareness of or interest in the object as such. But then an altogether new process begins: the process of valuation.

The content of volition and feeling assumes the form of the object. This object now confronts the subject with a certain degree of independence, surrendering or refusing itself, presenting conditions for its acquisition, placed by his original capricious choice in a law-governed realm of necessary occurrences and restrictions. It is completely irrelevant here that the contents of these forms of objectivity are not the same for all subjects. If we assumed that all human beings evaluated objects in exactly the same way, this would not increase the degree of objectivity beyond that which exists in an individual case; for if any object is valued rather than simply satisfying desire it stands at an objective distance from us that is established by real obstacles and necessary struggles, by gain and loss, by considerations of advantage and by prices.[9]

To put this argument from Simmel rather too simply: desire always depends on a distance from the object, and on the object's power to place the subject in the state of 'volition' – which is to reverse the general implication of the word volition, traditionally implying the agency of the subject – but only at a certain point does capital, value or money enter the equation as the intermediary between object and subject. Money appears – like the mobile apps of today – with the promise that it can connect us with what we want. We could take this suggestion in the following way today: is it possible to divorce desire from capital and create a different relationship between subject and its object, or are we now stuck in a world where desire is irrevocably tied to money?

A serious problem here is that the digital infrastructures of our world are so closely embedded in capitalism and its patterns that to use them at all is to be complicit. A critique of an argument I made that videogames could be repurposed for the Left was made by Jan de Vos. Countering what might be seen as a call to arms for the political Left to engage more directly in various levels of game production in my book *The Playstation Dreamworld*, de Vos argues that rather than inventing new gaming forms that might in various ways serve a progressive political agenda, we might be better served instead by seeking to question or rework the very issue or form of digital gaming itself, or even, perhaps, by addressing 'the forms of the digital technologies that underpin gaming as such'. In other words, 'game-dreaming about the leftist revolution is not likely to lead to any significant action' when the digital itself (on which videogames cannot help but rely) is so connected – as de Vos shows – to mainstream psychological conceptions of the human being and to a particular socio-economic political system: namely, that of capital. The same would be true of the love industries, mobile apps, and any kind of dating site. To a certain extent these are already part of the money economies that Simmel describes, whatever we might do with them

This raises the question of whether the potential technological solutions proposed in this book are themselves examples of game-dreaming about the leftist revolution, unless they also address the digital economy from which they arise. In other words, can we divorce the digital from capitalism? One significant issue here is that of hardware and the production line. Lewis Gordon took apart a PlayStation 4 to discover various materials such as plastics like acrylonitrile butadiene styrene (ABS) and polyoxymethylene (POM) and metals including gold, tin and enormous amounts of steel, showing the environmental impact of the console's hardware. He also traced manufactures of console parts to companies like Foxconn, Maintek and Casetek, many of which have been famed for wage circumvention and poor working conditions.[10] Gamesindustry.biz combed Sony's documents to discover more than 50 smelters or refiners who failed to conform to the Responsible Minerals Assurance Process.[11] An early advocate for focussing on hardware and production line of consoles is the games writer Marijam Didzgalvte, who has begun the work of thinking about how to reverse these patterns and produce an ethical gaming console.[12] If the digital proposals for a Left put forward in this book are to become serious projects, they will likewise have to source their parts differently.

What this shows is the huge capitalist network of 'analog' parts or raw materials upon which the existence of a digital object relies, but it doesn't approach the question of how desire might be divorced from that system. How can we move away from the SMV of The Red Pill Women and the social credit scores of programs like Baihe? Would it be possible to do so within a digital world so closely tied to the transnational global networks of capitalism?

SEXTOLOGY: DEVIANCE AND THE IMAGE BOARDS

Subversive acts of misogyny and deviance that have been in the spotlight in the last few years of the internet also seem to have something in common with the capitalist logic of our relationship to objects. In a study from 2011 – the days of Blackberry Messenger – Rosalind Gill and others identified the gender dynamics of 'sexting', showing that boys accumulated 'ratings' by possessing and exchanging images of girls' breasts, which operated as a form of currency and value, making this form of exchange both a prototype of the SMV system and a more mainstream and widespread example of its logic. They also illustrated how a 'blurring between pleasurable and coercive dimensions of digital sexual communication' characterised the sexting experience of young people.[13] They note that the process of sexting and digital image sharing can often be about systems of value where – generally speaking – male, white and economically privileged subjects are 'valued' where female, non-white and less privileged subjects risk devaluation. Generally speaking, we might add that while often solicited and sometimes even coerced digital images of young women enter into a marketplace of judgement, unsolicited and often unwanted 'dick pics' operate a much more accepted and less microscopically judged part of digital culture, appearing on screen – as it were – with a confident demand on the subject to respond either with desire or revulsion, with neither response experienced as inadequate by the sender.

Here we might apply the psychoanalytic concept of perversion to these practices: like the unwanted 'dick pic', the pervert calls the recipient into a particular subject position coded in advance by the act of perversion. What is theoretically important here is that the dick pic functions in the same way as the digital objects discussed above – from new cars to Pokémon – by forcing the subject to experience their own 'volition' and placing them in the position of desiring subject, even if revulsion remains possible. When the Google suggestion of a new latte or a Wish.com pair of saucy handcuffs appears uninvited into your newsfeed, the subject must act in response to the demand of assumed possible desire. The digital objects of consumer capital, then, function like an unsolicited dick pic and so it may be contemporary capitalism that is the truest pervert.

Not all forms of significant digital deviance stop at being misogynist and capitalist and the appraisal of the relationship between tech spaces and capital ought to be combined with attention not only to gender inequalities and to ones anchored in heteronormative ideas of sexuality but also to racial and intersectional biases. The most significant work here is that of Kishonna L. Gray, who has been appraising both the inherent biases of these spaces and their propensity for deviance and progressive reform for the last five years. She has shown that some of the most important examples of non-conformative or deviant behaviour within online social spaces connected to videogames are associated with homophobia and other forms of toxic hate speech. Practices like 'griefing' and 'salt-mining', well-known terms in multiplayer gaming discussion boards, refer to entering game spaces to send abuse and 'harvest' tears (salt) and trauma (grief), often publishing the evidence on image boards and forums. Another category is 'flaming', which 'can be understood as the spontaneous creation of homophobic, racist, and misogynist language during electronic communication', a definition which Gray notes is useful because 'of the incorporation of the word "spontaneous"', given that 'most of the hostile speech is unwarranted and comes out of nowhere'.[14]

This gives us two concerns. First, it shows that in dealing with the radical potential of subversive acts within these spaces we are up against not only the fact that the space is itself structured to limit or prohibit such acts but that those who do operate deviantly in the community are often those on the Right, with all those practices and image boards discussed above now associated with 'alt-Right' activism. Second, it shows that we need to be attentive to the way in which the digital space – as well as those acting within it – set the terms of the subject positions experienced on the platform. In her book, Gray discusses this in terms of Pierre Bourdieu's concept of habitus,

> For example, whiteness and masculinity are constantly reinforced in society, in schools, in government institutions, in the workplace, in TV, news, movies, and other media. It is now being reinforced in virtual communities.[15]

Whiteness and masculinity join capitalism as a set of values reinforced endlessly in the digital sphere. Like the unwanted dick pic, the salt-mining act of homophobia or of racism enters the digital sphere and demands to solicit a response. Like the coerced image of a young woman's body to be shared with school mates, the responses are submitted to forums and placed into a system of value where they are liked and upvoted, commented on and discussed. They are both approved by the forum onto which they are reported and approved again by virtue of their ability to create revulsion, trauma or disgust in the community outside of the forum.

Making this situation – which, as we have seen, runs from high schools to chan boards – possible is a particular form of digital organisation that has been identified in this chapter as the libidinal splinternet. By moving into micro-economies whose value systems inherit specific (often masculine, racially charged and economically driven) structures, users are able to experience pockets of extreme discourse as if that discourse is universal, trading in acts of aggression for credits in the miniature economy of approval on forums, image boards and private messaging groups. What we see here is that both mainstream culture (as with the use of Zhima Credit score to curate a national dating application) and 'deviant' communities (such as the salt-miners on 4chan) operate according to this logic of a particular kind of digital market value.

This is essentially the structure of image boards. Image boards occupy a strange place to the edge of mainstream internet culture and while sites like Reddit have a historical connection with liberalism and internet freedom campaigns, sites like 4Chan and 8Chan have at least superficially retained these values but have functioned to move discourse to the political Right and incubate extreme political activism. Structurally speaking, the most popular image boards follow a model that comes from the 'Futaba Channel', an early software which curates content into hierarchical subsections (opposed to this are 'Danbooru-style' image boards which host content non-hierarchically). Within these subsections, usually based on a topic or theme, users upvote, downvote and comment on images, providing each with a value and by proxy devaluing other images on the board. It is no surprise that far-Right image boards have often advocated for 'ethnopluralism' – the idea that each ethnicity might have its own place and space – given how they police the borders of these online communities. What may be more frightening about this is how the much more mainstream liberal internet also functions to organise people into subsets along classist lines with strong borders at their limits.

PITCH – MATCHTONOMY:
A DATING SIMULATOR FOR CLASS SOLIDARITY

In her polemical feminist text, Shulmaith Firestone presents the idea that when it comes to love women cannot afford to be spontaneous. That, she argues, if often the privilege of those in structural positions of power, in the context of her argument: the position of men. This is important to remember when it comes to the discourses that surround dating applications. Those who criticise the world of online dating usually argue that offline or 'traditional' dating offers a more organic, natural and spontaneous way of meeting a partner. Firestone's point raises the question of whether that would be in any way desirable, especially in a world in which economic conditions and precarity make every second of spare time precious and especially given those traditional structures of dating have always included inequalities.

On top of this, it's not at all certain that offline or pre-internet relationships are any more organic and spontaneous than the algorithm-driven ones we experience today. In *The Purchase of Intimacy* Viviana Zelizer discusses the kinds of legal and social contracts which have long since functioned to organise and sort people into couples:

> Outside the legal arena, in ordinary, everyday practice, people engage in a similar sorting of couples. They do not employ precisely the same distinctions as lawyers or invoke exactly the same moral evaluations of different kinds of relations. But they sort across the whole range of relations that involve the possibility of intimacy, from lawyer-client or doctor-patient to friends, neighbours, workmates, and kin.[16]

Social life functions to sort individuals into couples along economic, social and political lines. As we saw in the first chapter, dating applications inherit some of these social biases and embed them into algorithms whose data sets then go on to set the terms for friendships and relationships of the future. In light of the fact that it is perhaps both inevitable (Zelizer) and desirable (Firestone) to organise the political processes by which coupling takes place, we might playfully suggest the blueprint for an online dating app which might organise couples in more politically and economically desirable ways than the existing set of applications discussed in this chapter.

To make such a program, we'd need to think both about its algorithms and its interface. When it comes to the algorithm – the basic way it matches users with each other – we could start by describing the existing grouping structures used by typical matching algorithms and by thinking about how we might want to structure these processes differently. Since both Roland Barthes and Sigmund Freud have been used regularly in this book, we could playfully use the distinction that Barthes makes between 'metaphor' and 'metonymy', which relate to Freud's earlier ideas of 'condensation' and 'displacement', to differentiate the way dating apps and algorithms are from the way in which we might want them to be. As things stand now, we can say that dating apps work by a process of metaphor.

Metaphor and metonymy have often been considered two fundamentally different ways of humans structuring their discourses and ways of thinking. With metaphor, there is an impression of similarity (hence the long-standing association of metaphor with simile in the classroom) between the two items or objects in the process which are directly compared, as in Romeo's 'Juliet is the sun!' or in the world's most famous metaphor, 'shall I compare thee to a summer's day?'. In effect, this logic of similarity is the basis for dating sites and their algorithms, as well as – though perhaps to a lesser extent – for friend suggestions on social media networks. The logic of such algorithms is to organise people into large sets and connect them with individuals whose data is similar. Even in the interface of dating platforms, the logic is fairly clear: we select categories from a drop-down menu or answer extensive multiple-choice questionnaires which puts us into a pool with other users who selected these answers. On Hinge you can identify your politics by picking between liberal, moderate, conservative (notably omitting any leftist position), while on Bumble you can identify as liberal, Right, moderate or apolitical. Other questions are less directly political, such as OkCupid's 'Do you like the taste of beer?' or 'Would you like to be the Supreme Ruler of Earth?' The OkCupid official blog explains the process, saying 'if you answer "no" to the question "Should the government defund Planned Parenthood?" wouldn't you want your potential date to answer the same way?'. The algorithm works by finding matches with the closest similarity to you, seeing the ideal match as a kind of reflection of you (like the logic of the AI app Replika discussed earlier). In other words, there is a metaphorical logic of direct comparison between the two objects/lovers.

Opposed to this logic of connecting two things, Barthes discusses the idea of metonymy. Metonymy is often used casually to mean the substitution of a whole with a part – 'the part taken for the whole' as when the word 'suit' is used to describe a businessman. Rather than being based in similarity, the logic here is that there is an association between the two things. Another word for metonymy is contiguity, meaning to border upon something or to touch something. Lacan adds to this discussion in his famous essay 'The Instance of the Letter in the Unconscious' where he reconnects metaphor and metonymy with psychoanalysis. Here he suggests that metonymy has a more important connection to language, saying that 'the connection between ship and sail is nowhere other than in the signifier' and that therefore metonymy 'is based on the *word-to-word* nature of this connection'.[17] In other words, metonymy is more like the children's word association game where players must shout out the first word that comes to mind when another word is spoken: it is not so much about a similarity between the two words but about an association that exists in the unconscious that is made visible by the effects of language.

This could form a blueprint for designing an algorithm for a dating site that was able to cut through the kinds of social organisation that is common in existing sites. It might create a way of connecting people that uses data to make helpful suggestions while supporting a diverse range of class and cultural connections that could reach outside of the filter bubbles produced by the splinternet of cyberbalkanisation. If suggestions were based on metonymic connections between language used rather than similarity between users, all the data input into the site by the user would be taken into account and deployed, but the process would not preclude connections with those dissimilar to themselves. This might lead to occasionally unhelpful suggestions. For instance, if the word is 'left-wing' it would be proximate metonymically to 'right-wing' even though those two users might be unlikely to match, making the site rather the opposite of Trump.dating. However, since the algorithm would use hundreds of words rather than one, such a connection would be unlikely. Rather, users would be connected to those speaking in related terms to themselves. Wordmaps have been trending on Facebook and Instagram for the last several years as a means of reflecting on user's discourse, and such maps could form the basis of a data set that would not seek to connect the user with others using the same words but with related words in a huge big data map – a kind of word association game with millions of players – and aggregated into a model for making connections between people. In such a model, you would not be connected with those like you but those whose discourse connects to or intersects with yours.

In terms of how the interface might work, let's imagine a dating site structured like Reddit or like Wikipedia – a platform which works very much like the word association game (just think of the hours spent clicking thorough the rabbit holes of different pages and links). Such an experience would be almost the complete opposite to a swiping platform like Tinder or Hinge. Importantly, both are game-like, but they are completely different kinds of game. In one, you are a kind of digital detective, an amateur sleuth going through these links and holes, searching for connections and a place to rest your attention for a time, actively pursuing links that might be more likely to lead to that outcome. In the other, you are entirely passive and powerless as to which page will appear next, beholden only to the logic of the application. Each page on the site could be hosted by a single user, equivalent to their profile page on a dating site, and every word used on that page could be hyperlinked to other pages that have used that word or words metonymically connected to it, so that the user can jump from one profile to another by identifying a word or cluster of words of interest. For the most part, this would keep users in loops and bubbles of users who they share interests and connections with, but it would not preclude the user from actively moving outside of that bubble for any distance if they chose to pursue a particular line of interest or enquiry. Like with an hour spent on Wikipedia, users will find themselves far from where they usually expect to be, crossing the boundaries of the splinternet and its organisational capitalist logic.

For the most part this is a playful rather than a serious suggestion, but it hopefully at least points to the need for solutions to the problems of connecting in a data-driven society such as ours. It's clear that we need data and, as Firestone wrote, cannot always afford to be spontaneous. However, we also need ways to ensure we are not reduced to manageable bubbles of individuals partitioned off from each other by algorithms and curation tools which work by organising us according to class, race and gender demographics. If we are to foster greater solidarity with each other across cultural, economic and social borders, we need first to be able to see each other, and today that means appearing in each other's digital worlds.

Conclusion: Ready Worker One

The man who wants to dominate his peers calls the android machine into being

– Gilbert Simondon

In *The Twittering Machine* Richard Seymour writes that the social industries of digital media have created a machine for us to write to. For Seymour, 'the bait is that we are interactive with other people', whether that is our friends, professional colleagues, celebrities, politicians, royals, terrorists, porn actors – anyone we like. But 'we are not interactive with them, however, but with the machine'.[1] The idea has resonances of Slavoj Žižek's well-known discussions of the Lacanian Big Other: we send tweets out into the ether, less to each other but to an imaginary force of approval that comes to us as if from out of space. This god-like machine, writes Seymour, 'collects and aggregates our desires and fantasies, segments them by market and demographic, and sells them back to us as commoditiesw'.[2] There is now a very real force at the centre of social life organising our thoughts and behaviours and to whom we address ourselves daily. Ultimately, 'the internet will tell you who you are and what your destiny is through arithmetic "likes", "shares" and "comments"'.[3]

Is this yet another form of game, or at least of play, that characterises the digital space today? How far away are these likes, shares and comments from being the ultimate form of in-game reward, issued from the ultimate game engine of Twitter itself? As Seymour notes via Byung-Chul Han's idea of the gamification of capitalism, the addictive nature of social media approval feedback loops is not dissimilar to the effect of poker machines or smartphone games. Perhaps they are even alike to the endorphin-inducing chink of Sonic collecting a ring or of those candies crushing in a point-harvesting ping on Candy Crush. Our lives have become a game, with the rewards dished out by capitalism.

Roisin Kiberd's memoir of her life online testifies to how this situation permeates into our psyche and becomes part of the way we relate to other and part of the way we love. We often think we are engaging with others, when we are in fact engaging just as much with the strange and powerful machine.

> There's a danger in gauging someone's feelings for you by their online behaviour, because it might not be love that keeps us typing. It might be mutual boredom, or loneliness. It might even be the platform itself, because apps are engineered to keep us using them. What if I'm addicted to the medium, and not the message? What if, over all these years, I've been in love with Gmail, or Twitter, or Facebook, and the version of myself these platforms allow me to present?[4]

Love and desire are so implicated in the mediums we experience them through that it is hard to know whether it is our date or our friend that we desire, or the interface of Bumble itself with its game-like reward of a ping in our inbox, the approval of likes and re-tweets on Twitter or the reactions to a Facebook post.

This point has been made by others too. In Charlie Brooker's 'How Videogames Changed the World', first aired in 2014, many interviewees make the point that social media and social life have been infiltrated by games, with Twitter one obvious example. The emphasis there, as is often the case when this topic is discussed, is on how 'accidental' this process has been.[5] Games, or so the argument goes, have accidentally been invited into our social lives, which are now more game-like than ever. In this book, we have seen the love industries – connected to our deepest sexual desires and most intimate feelings of love and affection – are likewise gamified. But we have also seen that things are not as accidental as they might seem.

Twitter is not uniquely the problem, of course, but perhaps it does most perfectly embody the idea of digital social life today as a form of play and reward focussed around a centre that is both real and imaginary. Of course, its not the case that Mark Zuckerberg and Elon Musk were sat on the beanbags in their boardrooms and decided to instigate a mass social reorganisation at the level of desire to suit their businesses. What is the case is that the patterns of nascent platform capitalism, with unequal distribution of power at its heart, are developing – through no coincidence – in tandem with the new libidinal economy of desire.

This centralisation of power would ultimately need to be broken up for truly significant change to take place. At the moment, we have a concentration of power in the hands of those who set the terms of the game – largely Silicon Valley capitalists and their financiers – who can be thought of not so much as designers, for it is usually those under their employ who do the designing, but perhaps as architects of these black boxes of machine that run out social lives today. Everyone else takes up the role of a player. To play, then, is to be a subject of a particular kind of capitalism, one where inequalities of power are perhaps greater than ever. In this way, we really are all players, all gamers.

Even in a direct sense too, we are almost all gamers. More than 50 per cent of the world now plays videogames of some kind, so of course gaming is no longer the niche demographic it used to be when being a 'gamer' was considered a subcultural niche. In addition to that, gaming trends are, for the most part, not something exclusive to certain countries, even if they are of course implicated in the 'digital divide'. The African games market, for example, has exploded through fast developments in mobile gaming over the past decade. Gaming is also no longer an industry clearly demarcated along gender lines. Around half of games are played by women, and although there are still vast differences in the types of games men and women play – as we've seen in this book – gender politics are well and truly present along with those of race and class.

Things get more interesting in this regard when we think about the role played by gamification. With the technologies discussed in this book, we can say that even if a subject never plays games directly, gamification means that a lot of what they do engage with is part of the games industry in many ways. For instance, if a user is on a dating app like Tinder but does not consider themselves to actively be a player of games, they are still being influenced by gamified elements because the swiping of Tinder is clearly a form of gamification, in that gamification most broadly means when elements from the gaming industry make their presence felt in other areas of life. In the case of the gamification of the dating world, this is even backed up by corporate patterns of ownership, as we've seen through games company Beijing Kunlun Tech's acquisition of Grindr. Similarly, we've seen that this exists with in-game rewards for various kinds of actions in the city such as those considered by Transport for London to offer in-game rewards to remove route congestion or those employed by the infamous Beijing Social Credit system or by the Alibaba Zhima Credit score used by the Baihe app. In this way games, their technologies and their 'playability' are fundamentally linked to the future of how our society is being constructed. We might even consider that ordering from Deliveroo is quite a gamified experience, albeit in a more abstract way. The point here is that, whether we play games or not, we are all affected by these trends. To be a gamer then, is simply to be subject to this power which operates on us and through us at the very level of our desire.

THEORIES OF DESIRE

At the beginning of this book it was proposed that the smart city of today is a city of desire, desire plugged into corporations and governments via the artificial limbs of our smartphones, tablets and game consoles. In many ways each citizen lives a highly individualised life of their own desires which can apparently be satisfied individually and uniquely by personalised algorithms pitched specifically to us which we become dependent upon and invested in. There is, of course, a long history of seeing desire in this way: as *belonging* to the individual and as the subject's own responsibility. This assumption is something psychoanalysis has specifically sought to dispel. Rather than originating simply from within, psychoanalysis has stressed that desires operate on the boundaries of the subject and their external world, often appearing as instinctual when they in fact come from external social life. Lacan makes a distinction between 'drive' and 'instinct' to stress the point that sometimes we think we are acting on internal impulse when we are in fact driven from outside.

> *Trieb* [drive] gives you a kick in the arse, my friends – quite different from so-called *instinct*. That's how psycho-analytic teaching is passed on.[6]

We are nudged to do something or to want something, but we feel as if we have chosen to do it ourselves or that we want it instinctually. We are moved to do something not so often from within but from a kick in the arse that drives us in a certain direction (quite literally in the case of the predictive Alibaba car). In exactly this way, when we act in the libidinal economy of the digital city today, we are often driven by forces outside of ourselves rather than in accordance with our own 'libido'.

Media theorist Dominic Pettman's *Peak Libido* dedicates a chapter, 'Whose Libido?' to the complex question of the origins of desire in different conceptions of psychoanalysis. Pettman deals with the most significant critics of psychoanalysis in Deleuze and Guattari, whose notions of the 'desiring machine' and of the 'dividual' are often the lens of choice for conceptions of the digital subject. Deleuze and Guattari's critiques of psychoanalysis are made over the two-volume *Capitalism and Schizophrenia*, consisting of *Anti-Oedipus* and *A Thousand Plateaus*. There they argue that psychoanalysis is guilty of tying its ideas of desire too closely to the individual human subject, taking particular issue with Lacanian ideas that desire is predicated on a lack (the classic psychoanalytic idea of subjectivity as fundamentally an experience of lack) at the centre of identity, which they believe falsely posits the subject as that which comes first for desire to follow after as something that only exists in relation to the established psychoanalytic subject. For Lacan, from his famous 'mirror stage' essay onwards, the subject is fundamentally constructed via a sense of lack in relation to its own image. This is metaphorically (and playfully) imagined as beginning from the moment the child first sees its reflection, from which point on the subject begins to experience desire: desire to redress this perception of lack and inadequacy.

On the contrary, Deleuze and Guattari suggest the concept of the dividual, discussed in Chapter 3, a subject who is multiply divided and always-already dividing along the lines of desire. For them, this idea of the subject – which can certainly be seen in contemporary consumer and digital capitalism with its endless insistence on the pursuit of many pleasures – is something of an antithesis to the more structured desiring subject envisioned in psychoanalysis. Following this critique, Pettman discusses how psychoanalysis 'was incapable of seeing desire as non-goal-directed' and explores the ideas of Alphono Lingis, whose strange book on desire *Excesses* argues that libido should be seen precisely as 'desire without being desire for something'. For Lingis, the 'sex urge is not naturally a craving of a male for a female; it is culture, repression and taboos that will narrow it down to a certain sex object'. For Pettman, psychoanalysis itself is part of the cultural process which narrows desire down to a goal-oriented desire for something.[7]

This presents a major debate about the relationship between psychoanalysis and media theory, and in particular between those advocates of Deleuzian theory and those of psychoanalysis. Although the approach in this book has been directly psychoanalytic, some of its arguments are closer to Deleuzian models of desire than is often the case with psychoanalytic approaches, given the assumption of irrevocable tension between the approaches by many of those who deploy them. The idea of the 'dividual' is a useful tool in thinking about the citizen of the smart city discussed in Chapter 2, for example. This kind of divided libidinal subject is in some ways precisely the subject of Pokémon GO or of Alibaba's desire-anticipating smart car, who follow an infinite array of desires as they move around the city in line with the routes set out by technocapitalists. Likewise, the subject of the dating simulators discussed in Chapter 3, who is able to step from their own individual world of desire into the imaginary shoes of another desiring character, could be thought of in these terms.

However, what has also been a consistent feature of the arguments made in this book is that the libidinal technologies deployed in the media industries of today work so effectively on their subjects precisely because the subject experiences their pleasures *as if* they are their own. The citizen of the technological spaces discussed here relates to their drives as if they are instincts, to use the Lacanian terms. In this way, while in some ways our desires are as divided as Deleuze and Guattari claim, psychoanalysis – a discourse primarily interested in the borders between the subject and the external world – remains vital in exploring how the subject experiences and relates to these desires. Freud himself commented that with technology 'man has become a God with artificial limbs' and it is this psychological transformation that takes place with technology that radically transforms what it means to be a human today. In short, what psychoanalysis offers is a way to make visible how we are being tricked at a psychic level by the economies of desire found in capitalism.

Another potentially incongruous argument to the psychoanalytic ideas of desire in contemporary capitalism proposed here that Pettman considers is via Steigler's concept of 'peak libido', from which Pettman gets the title of his book. Pettman outlines Steigler's position in the language of drives discussed here:

> Libido has been solicited to the point of scarcity. It is unsustainable, fracked almost out of existence by technologies that instead need only a kind of minimal zombified 'drive' in order to create profits. 'Peak libido' thus signals a situation where the most essential human resource of all – the libido, the life force itself, which seeks to foster a future with other human beings (and other inanimate allies) – is rapidly being depleted.[8]

For Steigler, we can no longer depend on a 'structure of feeling' (as psychoanalysis sometimes might) because of the 'disorienting, atomizing effects of modern technology'.[9] Pettman situates this situation in relation to Japan, famous for its hyper-sexualised and apparently obsessive society that produces products like the Gatebox and many of the simulators discussed in this book, and in relation to the US, showing that sex has radically declined in recent years even in those countries where it might seem to be so prominent culturally. In this situation of peak libido, we are left with little desire, undertaking the endless expenditure of capital in small zombified gestures of unfulfilling, micro-pleasure-yielding moments that characterise contemporary life.

This absence of libido can be the cause of alienation, depression and perhaps even of acts of extremist outburst and rebellion, from school shootings to other forms of mass murder.[10] From this perspective, it is the lack of desire itself which is the cause of dangerous symptoms like acts of violence. Here we might find an answer to the long-standing question of how videogames relate to such acts, given the endless media connection made between the world of games and school shootings. Many (including of course Donald Trump, in an attempt to turn the conversation away from gun laws) have argued that games – by offering simulations of violence – might encourage the replication of those acts in real life, where others have argued that the simulation would opt as an outlet for violence and thus protect the 'real' world from such expressions. Earlier we saw these arguments repeated in public debates around sex robots, noting that both positions risk naturalising the act of violence itself. In light of these arguments, it might be that rather than having a cause and effect relationship, both the desire for simulation and the impulse to extreme acts of violence are both (very different) symptoms of an absence of desire. While one offers the user the chance to experience the simulated desire of their other to cover the lack of their own (see Chapter 3), the latter represents a violent burst in an attempt to produce something to desire or to act *as if* these extreme desires existed.

The idea of 'peak libido', then, has as many parallels with Lacanian models of desire in the field of psychoanalysis as it does tensions. In the world of psychoanalysis, at least since Lacan, desire for an object, person or thing is always seen as a copy or even simulation of an earlier desire in the subject's psychological past which has been given up or barred in the subject's development. This earlier desire, a kind of pure intensity of libido, is given the famous name *object a* in Lacan's work. It is something that drives the subject and compels them to act, something that desire tends towards and reaches for. The endless and multiple objects of desire that the subject experiences a draw towards – from moments on PornHub to Pokémon and pokebowls – operate as stand-in replacements that the subject switches between in its psychic life. Of course, the more it switches from ultimately unfulfilling object to ultimately unfulfilling object, the more it drives profit for consumer capitalism.

Lacan puts important emphasis on the fact that experiencing the pure chaotic libido associated with the *object a*, actually getting 'what you want' would be a completely undesirable and dangerously traumatic process. In other words, replacement desire or simulated desire is more manageable and even more 'desirable' than a situation (completely imaginary in any case) in which a subject could in fact access this pure original desire. Though the point is different, this makes the idea rather opposed to the implications of a 'peak libido' argument which laments a lack of active desire in contemporary digital culture. From a psychoanalytic perspective, we could say that all subjects exist in a state of secondary desire, not pursuing primary objects of desire but 'zombified' replacements, and that this is not a negotiable situation but a very condition of being a human subject.

This provides us with two major takeaways. First, it demands that discourses of desire move away from ideas of cultural, social and economic conditions being structured around giving subjects what they want. This would fundamentally cut against a major strand in capitalist rhetoric that is increasingly prominent in tech communities from digital marketing and advertising to personalisation and curation strategies to social media and data collection mechanisms. These industries endlessly claim to be geared towards giving the subject what it wants, when we have consistently seen that their technologies are deployed much more in transforming what the subject wants than offering up objects to an already existing desire. In this sense, technocapitalists are already acting in a strangely psychoanalytic way, working on the ways in which citizens' displaced desires can be ordered and reordered to suit their agendas. At the same time, they utilise the capitalist narrative of offering their citizens what they want to obscure this fact. On the contrary, progressive political voices seem not to sufficiently accept this situation: that we need not so much to give citizens what they want or to liberate their desires as to enact a revolution in desire that transforms the libidinal economy of contemporary social life.

Second, it points to how potentially dangerous existing models of desire circulating in contemporary capitalism can be. A consistent rhetoric of intense pursuit of individual desires combined with a situation of peak libido intensifies the forms of depression and feelings of unfulfillment which seem to characterise contemporary digital life, leading to a situation where desires must be acted on or responded to rather than transformed. Further, this whole model of desire as emanating from within the individual and existing before or prior to the subject's immersion into the technologies of digital capitalism instils the idea that desires are individual rather than collective. The whole infrastructure of unique personalised digital marketing and page curation feeds this imaginary world of naturalised desire in the capitalist imagination. It is not so much that desires are naturally or instinctively individual and that the mechanisms of capitalism answer these desires, but that these technologies have manipulated desires into more individualised and unique experiences to suit a capitalist agenda. A collective progressive imagination would need to reorient desire for mutual collective benefit to reproduce collective desire. Something that united the playful suggestions at the end of each chapter in this book is that each proposal attempts to build a pleasurable, gamified experience of contemporary life that is collective in its approach.

Of course, to make real change, we'd need to do more than propose some ways in which – were the power relations in tech ownership different – we could use new technologies to promote socialist causes, as some of the suggestions in this book have done. We would also need to transform a situation in which we are all players in a gamespace over whose architecture we have little or no power. In their excellent book *Work Want Work*, Mareile Pfennebecker and James Smith conclude that significant structural change is the only way out of what they call an 'affective manipulation' by corporations:

> To win the struggle for cooperative or public non-profit platforms, as it is increasingly demanded by the digital left, would be the first step: towards writing better code for our social lives online that, instead of keeping us there at the cost of addictive affective manipulation and reliably simplified pleasures, leaves more room for chance, or breaks, for the kind of surprise encounters currently edited out of our search results as well as our sexual preferences. Of course, these new platforms could only show their full potential in a world where there is

sufficient time outside work to pursue their pleasures, but their possibility demonstrates the point; if we do not want digital capitalism to put our desires to work, structural change, not individual retraining, is required.[11]

Whether spontaneity and surprise are what we want (Firestone might disagree), the point here is that if we do not want desire to work solely for digital capitalism, we will need structural change and not just some quirky suggestions for how we might use technology for the Left. Platforms and their technologies need to be cooperative, not-for-profit, open access and readily available if they are to be repurposed and re-used in creative ways to serve a different desire agenda to the one they are serving today. Again though, we need to be wary of the language of corporate capitalism and its own logic of giving out things for free.

FREE STUFF

These arguments are well known in media studies. The names most closely associated with the discussions are Richard Stallman and Lawrence Lessig. In the early 1980s, Stallman founded the Free Software Movement to advocate for an opening up of the digital sphere and to combat its ownership models and their neoliberal ownership ethic. For Stallman, free software has little to do with being 'free' in the sense that the user does not have to pay for it. To be 'free' in this context meant to be non-proprietary, to have its source code open and available and to have transparency so that users could see how the software functions and operates. These campaigns for Free Software developed over the decade and into the years of the early internet into a radical challenge to digital rights and a force in the struggle for cooperative or public nonprofit platforms. In the late 1990s though, the discourse around free software also began to be harnessed by businesses and big media for its own agenda.

In *TwoBits* Christopher Kelty outlines the divide that took place by comparing 'free software' models based on Stallman's activism with 'open source' technologies (which Stallman himself criticised)[12] that began to be advocated by businesses in the dotcom boom:

Free Software forked in 1998 when the term *Open Source* suddenly appeared (a term previously used only by the CIA to refer to unclassified sources of intelligence). The two terms resulted in two separate kinds of narratives: the first, regarding Free Software, stretched back into the 1980s, promoting software freedom and resistance to proprietary software 'hoarding', as Richard Stallman, the head of the Free Software Foundation, refers to it; the second, regarding Open Source, was associated with the dotcom boom and the evangelism of the libertarian pro-business hacker Eric Raymond, who focused on the economic value and cost savings that Open Source Software represented including the pragmatic (and polymathic) approach that governed the everyday use of Free Software in some of the largest online start-ups (Amazon, Yahoo!, HotWired, and others all 'promoted' Free Software by using it to run their shops).[13]

Open source software today is given out by all the major tech companies discussed in this book, and famous examples include the Google Maps API discussed in Chapter 2, which is available for free to developers who then become dependent on Google for their creative work and programs. In his book with the important title *Free Culture: How Big Media Uses Technology and the Law to Lock Down Culture and Control Creativity*, Lawrence Lessig developed the arguments of Stallman and argued for a stifling and restricting of creativity as a result of 'free software' and its proprietary control by an increasingly small number of powerful stakeholders. For Lessig, the internet's early liberatory and transformative potential has not only been quashed and controlled but turned completely on its head, leaving us in a situation where the internet removes freedoms and collectivist features that had previously been – at least to some extent – present.

> While the Internet has indeed produced something fantastic and new, our government, pushed by big media to respond to this 'something new', is destroying something very old. Rather than understanding the changes the Internet might permit, and rather than taking time to let 'common sense' resolve how best to respond, we are allowing those most threatened by the changes to use their power to change the law – and more importantly, to use their power to change something fundamental about who we have always been.[14]

Ownership rules and intellectual property laws have privatised the space of the internet even, and in fact particularly, when its products and services are most 'free'. This open source 'free' internet – an echo of neoliberal economic discourse – set the blueprint for the kinds of platform capitalism we see today. In the terms of our own argument, we have become players in the games of platform capitalism and each act of play generates content and data of value, value that is harvested only by that small group of stakeholders. Free platforms and open access have become primary mechanisms through which we are put to work by these small groups of powerholders to have our labour harvested and collected for profit. Projects like Laurent Ptak's 'Wages for Facebook' highlighted this fact, but alarm bells should ring about the idea of being paid per post when it was included in the proposals of Silicon Valley Democratic candidate Andrew Yang's 2020 election campaign. Rather, as Smith and Pfennebecker argue, we'd need a way to share the platform properly to have any hope of a collective digital future. What Lessig points out is the collective and democratic potential of the internet – shown in Stallman's free software project – has been bypassed in favour of an unprecedented shutdown of collective space.

Lessig also shows that this produces a kind of stifling of creativity, and the argument here is not simply a cultural one about the need for more open creative platforms but a technical one about how the economic structure of technology limits its potential advancement. In concrete ways, the dominance of the digital industries by big tech reduces rather than increases the amount of creative production that can happen at the forefront of technology. While Google, Facebook, Tencent, Alibaba, Tesla and other Silicon Valley-style corporations are universally heralded as the pioneers taking us into the technological future, it is in fact possible that these companies stifle rather than accelerate technological progress.

Many free software projects, not unlike some of those found on platforms like Github, were able to develop quickly and efficiently because of a 'hive-mind' of users whose free and open access to the software meant that they could contribute to testing, editing and improving digital technologies as they developed. This method of working on digital creation was often more advanced and efficient than those methods being employed in Silicon Valley, which tended to create beta products in-house and then outsource testing, accumulate feedback and then amend the product. Instead, the organic open access method allows for real-time 'organic' testing, development, feedback and redesign. What happened next is that major corporations like those discussed here worked hard to harness the power of these kinds of worker-driven development and turn them to their own profit, steering the internet away from Stallman's idea of a more open and democratic space for collaborative work.

In moving from 'free software' to 'open source', the discourses of neoliberal 'platform capitalism', nascent at this time, began to take hold of this space and territorialise it, allowing individuals to use and work in the space but ensuring its ownership remained private and the profits from its productivity were accruing only in 1 per cent of the pockets. As early users of this evolving consumer content-driven internet (which would later be termed Wed 2.0) began to enjoy and participate in the pleasures of content creation, design and improvement, technocapitalists began to see how libidinal and desire-driven the economies of the future would be and started to work to harness the creative power of the masses for its own profit. If these users could be freed from working for the 1 per cent, its possible they might produce considerably more of the future than Google. We need individual re-training, a collective hivemind of collaborative workers and policies and regulations that will wrest ownership from the 1 per cent, if we are to have a chance in this revolution.

THE FUTURE OF DESIRE

Web 3.0 is a term that has been given to the movement of artificial intelligence and automation into this space of the community internet. Web 2.0 is characterised by platforms like Facebook and even Uber, which are known as 'lean' platforms, meaning that users function as both workers and consumers and the platform (itself minimal in substance) collects the profit. Web 3.0 refers to a situation in which such platforms begin to bring artificial intelligence and algorithms in as mediators between people and things. Web 3.0 allows distributed users and machines to interact with data, value and other objects via peer-to-peer networks with human third parties replaced – essentially – by robots. While Silicon Valley pioneers have praised the development as human-centric and privacy preserving, in fact what we have consistently seen in the case studies in this book is that the handing over of mediation, communication and even thinking itself is tending to serve only the digital elite.

What we have also seen is that if there is a Web 3.0 then it has desire at its centre – desire as a resource or raw material which itself is grabbed and territorialised by a digital media industry with its own politics, agendas and ambitions. As Gilbert Simondon wrote, 'the man who wants to dominate his peers calls the android machine into being. The welcoming of AI automation into the internet could be little more than the next step in the power-grabbing of the 1 per cent. Eulogistic praise for the new internet and the autopia to come needs to be seriously tempered. From a socialist, progressive or even community-driven perspective, we should be very wary.

There is to be a new internet indeed. It is being built as we write and read, and the new terms of the future are being carved out. This internet works by exercising more and more control on its subjects, and what we learn from the digital love industries that are now a vital part of daily life is that it is our very desires through which we are becoming the subjects of the future.

From sex robots and smart condoms to dating sites, simulators, videogames and pornography, we are seeing a 'desirevolution' take place before our eyes. There will be no restoration when this revolution has been completed. Things will not go back to how they were. All that remains to be decided is who the victors will be in this battle for the future of desire (and therefore of politics). At the moment, progressive forces might be struggling to catch up with the populist Right and lagging miles behind the neoliberal capitalists, but the battle is not yet over. Love has been the site of this battle – a key to controlling and reorganising us into the perfect functioning capitalists of the smart city. Love can also be a way to fight back, to produce collectively and to desire collectively for a future with solidarity and a commons in which we can all live and love.

Notes

Websites last accessed 5 September 2021.

INTRODUCTION: THE GRINDR SAGA

1. Evan Moffitt, 'Under My Thumb' in *Frieze*, 208 (January–February 2020).
2. Gavin Brown, 'Being Xtra in Grindr City' in *How to Run a City Like Amazon and Other Fables*, ed. Mark Graham, Rob Kitchin, Shannon Mattern and Joe Shaw (Meatspace Press: London, 2019), para. 553.
3. Allison de Fren, 'Technofetishism and the Uncanny Desires of A.S.F.R. (alt. sex.fetish.robots)' in *Science Fiction Studies*, 36, pp. 404–40.
4. McKenzie Wark, *Gamer Theory 2.0*, 'Agony on The Cave', card 013, available at: www.futureofthebook.org/gamertheory2.0/
5. Bogna Konior, 'Determination from the Outside: Stigmata, Teledildonics and Remote Cybersex' in ŠUM, 12.
6. Ibid.
7. Solange Manche, 'Tinder, Destroyer of Cities – When Capital Abandons Sex', in *Strelka Mag* (20 September 2019).
8. Ian Parker and David Pavon-Cuellar, *Psychoanalysis & Revolution* (London: 1968 Press, 2021).
9. Anon, *Red Therapy* (London: Rye Express TU, 1978), p. 4.
10. Todd McGowan, *Capitalism and Desire* (New York: Columbia University Press, 2016), p. 7.
11. Herbert Marcuse, *Eros and Civilization: A Philosophical Inquiry into Freud* (New York: Routledge, 1987), p. 46.
12. McGowan, *Capitalism and Desire*, p. 21.

1 DATA LOVE

1. See Judith Deportail, 'Dans le laboratoire de la "fake science"' in *Le Temps* (7 April 2017).
2. Yuk Hui, *On the Existence of Digital Objects* (Minneapolis, MN: University of Minnesota Press, 2016), p. 50.
3. Alain Badiou, *In Praise of Love*, trans. Peter Bush (London: Serpent's Tale, 2012), p. 57.
4. Ibid., p. 60.
5. Ibid., pp. 62–3.
6. Eva Illouz, *Why Love Hurts: A Sociological Explanation* (London: Polity, 2012), p. 5.

7. Shulmaith Firestone, *The Dialectic of Sex: The Case for Feminist Revolution* (New York: Bantam Books, 1970), p. 126.

8. Ibid., p. 130.

9. Illouz, *Why Love Hurts*, p. 6.

10. See www.laboriacuboniks.net/; see also Helen Hester, *Xenofeminism* (London: Polity, 2018).

11. Asad Haider, *Mistaken Identity: Race and Class in the Age of Trump* (London: Verso, 2018), p. 81.

12. See Michel Foucault, *The History of Madness*, trans. Jonathan Murphy and Jean Khalfa (London: Routledge, 2006), pp. 44–77.

13. Henri Lefebvre, *Critique of Everyday Life: Volume Three* (London, Verso, 2005), p. 151.

14. Slavoj Žižek, *The Universal Exception*, ed. Rex Butler and Scott Stephens (London: Continuum, 2006), p. 152.

15. Ivan Chtcheglov, 'Formulary for a New Urbanism', via Bureau of Public Secrets, available at: www.bopsecrets.org/SI/Chtcheglov.htm

16. Dominic Pettman, *Infinite Distraction* (London: Polity, 2016), p. 9.

17. Hui, *On the Existence of Digital Objects*, p. 47.

18. Reprinted in Jeremy Tambling, *Re:Verse* (London: Routledge, 2007), p. 191.

19. See Alfie Bown, *The Playstation Dreamworld* (London: Polity, 2017).

20. Sheila Jasanoff, 'Future Imperfect: Science, Technology and the Imaginations of Modernity' in *Dreamscapes of Modernity: Sociotechnical Imaginaries and the Fabrication of Power*, ed. Sheila Jasanoff and Sang-Hyun Kim (Chicago, IL: University of Chicago Press, 2015), pp. 1–33 (p. 2).

21. Srećko Horvat, *The Radicality of Love* (London: Polity, 2015), pp. 1–23.

22. Roland Barthes, *A Lover's Discourse: Fragments*, trans. Richard Howard (New York: Farrar, Straus and Giroux, 2001), p. 192.

23. See Slavoj Žižek, *Like a Thief in Broad Daylight: Power in the Era of Post-Humanity* (London: Penguin, 2018), p. 33.

24. Johann Wolfgang von Goethe, *The Sorrows of Young Werther* (London: Penguin, 1989), p. 37.

25. Niklas Luhmann, *Love as Passion: The Codification of Love* (Cambridge, MA: Harvard University Press, 1986), p. 29.

26. Ibid., p. 45.

27. Dominic Pettman, *Look at the Bunny: Totem, Taboo, Technology* (Zero Books, 2013), p. 100.

28. Ibid., p. 118.

29. Laurent Berlant, *Love/Desire* (London: Punctum Press, 2012), pp. 6–7.

30. Sigmund Freud, 'Group Psychology and the Analysis of the Ego' in *The Standard Edition of the Complete Words of Sigmund Freud vol 18*, trans. James Strachey (London, Vintage: 2001) pp. 67–144 (p. 138).

31. Jean Baudrillard, *The Ecstasy of Communication* (Los Angeles: Semiotext(e), 2012), p. 20.

2 THE DIGITAL LIBIDINAL CITY

1. Ibid., p. 20.

2. Lee Grieveson, *Now: A Media History* (forthcoming).

3. Guy Debord, 'The Theory of the Dérive' in *Situationist International Anthology*, ed. Ken Knabb (Berkeley, CA: Bureau of Public Secrets, 1981), pp. 50–4.

4. For the full breadth of this argument see 'The Pokémon Generation' in Bown, *The Playstation Dreamworld*, pp. 1–27. Shoshana Zuboff discusses Pokémon GO as a key aspect of behavioural modification in the city two years later in 2019, but still neglects the libidinal nature of these changes. See Shoshana Zuboff, *The Age of Surveillance Capitalism* (London: Profile Books, 2019), pp. 3–18.

5. One context in which this has been demonstrated by research is in Dan M. Kotliar's article on three 'choice-inducing' Israeli applications. See 'Who Gets to Choose? On the Socio-algorithmic Construction of Choice' in *Science, Technology, & Human Values* 46(2) (May 2020).

6. Hui, *On the Existence of Digital Objects*, p. 1.

7. Cornelius Castoriadis, *The Imaginary Institution of Society*, trans. Kathleen Blamey (Cambridge, MA, 1987), pp. 288–90.

8. Bill Brown, 'Thing Theory' in *Critical Inquiry*, 28, No. 1, Things (Autumn, 2001), pp. 1–22 (p. 8).

9. Jacques Lacan, *On Feminine Sexuality: The Limits of Love and Knowledge Book XX*, trans. Bruce Fink (London: W.W. Norton, 1999), pp. 90–1.

10. Nick Srnicek, *Platform Capitalism* (London: Polity, 2018).

11. Christian Fuchs, 'Internet and Class Struggle', with Benjamin Burbaumer, available at: www.historicalmaterialism.org/node/963

12. Mario Tronti, *Workers and Capital* (London: Verso, 2019).

13. Benjamin Bratton, *The Stack: On Software and Sovereignty* (Cambridge, MA: MIT Press, 2015), p. 124.

14. Ibid., p. 127.

15. Italo Calvino, *Under the Jaguar Sun*, trans. William Weaver (London: Penguin, 2009), p. 27.

16. Ibid., p. 66.

17. Alenka Zupancic, *What Is Sex?* (Cambridge, MA: MIT Press, 2018).

18. Elinor Carmi, 'The Hidden Listeners: Regulating the Line from Telephone Operators to Content Moderators' in *International Journal of Communication*, 13 (2019), pp. 440–58.

19. Rob Horning, 'Preemptive personalization' in *The New Inquiry* (11 September 2014), available at: http:// thenewinquiry.com/ blogs/marginal- utility/ preemptive-personalization

20. Armen Avanessian, *The Speculative Time-Complex*, ed. Arman Avanessian and Suhail Malik (Miami: NAME Publications, 2016), p. 7.

21. Thanks here are due to James Smith, with whom a co-authored article on this subject was published which started the work of this section.

22. See 'The Data That Turned the World Upside Down', available at: https://motherboard.vice.com/en_us/article/mg9vvn/how-our-likes-helped-trump-win

23. See Angela Nagle, 'Enemies of the People' in *The Baffler*, No. 34 (March 2017).

24. Raymond Williams, *Resources of Hope: Culture, Democracy, Socialism* (London: Verso, 1989), pp. 3–14 (p. 11).

25. Will Davies, 'Leave, and Leave Again' in *London Review of Books*, Vol. 41, No. 3 (7 February 2019), pp. 9–10.

26. James Smith and Mareile Pffanebecker, 'What Will We Do in the Post-Work Utopia?' available at: http://blogs.lse.ac.uk/lsereviewofbooks/2016/06/17/the-long-read-what-will-we-do-in-the-post-work-utopia-by-mareile-pfannebecker-and-j-a-smith/

27. Sigmund Freud, *The Standard Edition of the Complete Psychological Works of Sigmund Freud, Volume XVIII (1920–1922): Beyond the Pleasure Principle, Group Psychology and Other Works*, trans. James Strachey (London: Vintage, 2001), pp. 65–144.

28. Freud, 'Group Psychology and the Analysis of the Ego', p. 138.

29. Jacques Lacan, ... *or Worse: The Seminar of Jacques Lacan* (London: Polity, 2018), pp. 191–2.

30. See Bown, *The Playstation Dreamworld*.

31. Samo Tomsic, *The Labour of Enjoyment: Towards a Critique of Libidinal Economy* (Berlin, August Verlag, 2019), p. 63.

32. Matt Goerzen, 'Notes Towards the Memes of Production', *Texte zur Kunst* 106, pp. 82–108.

33. Isabel Millar, 'Baudrillard: From the Self-Driving Car to the Ex-timacy of Communication?' in *Everyday Analysis* (February 2019).

34. Baudrillard, *The Ecstasy of Communication*, p. 57.

35. Jacques Lacan, *Seminar VII: The Ethics of Psychoanalysis*, trans. D. Porter (London: Norton, 1992), p. 324.

36. Cindy Zeiher, 'The Subject and the Act: A Necessary *Folie à Deux* to Think Politics' in *Filozofski vestnik*, 37 (2016), pp. 81–99 (p. 86).

37. Ibid., p. 98.

38. William Davies, *Nervous States: Democracy and the Decline of Reason* (London: Jonanthan Cape, 2018).

39. Georges Perec, 'Approaches to What?' in *Species of Spaces and Other Pieces*, trans. John Sturrock (Harmondsworth: Penguin, 1997), p. 205.

40. Michel de Certeau, *The Practice of Everyday Life*, trans. Steven Rendall (London: University of California Press, 1984), p. 92.

41. Dawn Lyon, *What Is Rhythmanalysis?* (London: Bloomsbury, 2019), p. 13.

42. Darian Leader, *Why Can't We Sleep?* (London: Penguin, 2019).

43. Phoebe Moore, 'The Quantified Self: What Counts in the Neoliberal Workplace' in *New Media & Society*, 18, No. 1 (2016), pp. 2774–92; Jingyi Wan, 'How Should We Behave in Class' in *Everyday Analysis*, available at: https://everydayanalysis.net/2020/02/07/how-should-we-behave-in-class/

44. Lefebvre, *Rhythmanalysis*, pp. 36–7.

45. Ernest Hartmann, *The Sleeping Pill* (London: Yale University Press, 1978), p. 131.

46. For the scope of this argument see Alfie Bown, *Enjoying It: Candy Crush and Capitalism* (Winchester and Washington: Zero Books, 2015), pp. 29–32.

47. Gayatri Chakravorty Spivak, *Death of a Discipline* (New York: Columbia University Press, 2003), p. 72.

48. Lukáš Likavčan, *Introduction to Comparative Planetology* (Moscow: Strelka Press, 2019), p. 18.

49. Ibid., p. 101.

50. Gayatri Chakravorty Spivak and Susanne M. Winterling, 'The Imperative to Make the Imagination Flexible' in Pandora's Box (2015), available at: http://pandorasbox.susannewinterling.com/imperative-make-imagination-flexible

51. Likavčan, *Introduction to Comparative Planetology*, p. 79.

3 SIMULATION AND STIMULATION: FROM GAMES TO PORN

1. Jeremy Paris, 'The Most Important Games on Sega's SG-1000' in *US Gamer* (26 December 2013).

2. See Bown, *The Playstation Dreamworld*, pp. 41–9.

3. See Andrew Pollack, 'Japan's Newest Young Heartthrobs Are Sexy, Talented and Virtual' in the *New York Times* (25 November 1996, Section D), p. 5.

4. See www.gatebox.ai/en/

5. Wark, *Gamer Theory 2.0*.

6. Pettman, *Look at the Bunny*, pp. 99–100.

7. Carolina Bandinelli and Arturo Bandinelli, 'What Does the App Want? A Psychoanalytic Interpretation of Dating Apps' Libidinal Economy' in *Psychoanalysis Culture & Society*, 28 (2021), pp. 181–98 (p. 193).

8. Oliver Grau, *Virtual Art: From Illusion to Immersion* (Cambridge, MA: MIT Press, 2003), pp. 5, 27.

9. Janet Murray, *Hamlet on the Holodeck: The Future of Narrative in Cyberspace* (Cambridge, MA: MIT Press, 1997), pp. 98–9.

10. Ibid., pp. 98–9.

11. Edwin Montoya Zorrilla, 'VR and the Empathy Machine' in *The Hong Kong Review of Books* (December 2016), available at: https://hkrbooks.com/2016/07/22/hkrb-essays-pokemon-go-and-the-enigma-of-the-city/

12. Charlotte Veaux and Yann Garreau, a conference delivered at *Society of Immersive Experiences*, 2019.

13. Maria Chatzichristodoulou, 'Immersed in Otherness', a talk given at *New Art Exchange*, Nottingham (February 2020).

14. See for example Feona Attwood, 'No Money Shot? Commerce, Pornography and New Sex Taste Cultures' in *Sexualities*, 10, No. 4 (2007), pp. 441–56 and Danielle DeVoss, 'Women's Porn Sites: Spaces of Fissure an Eruption or "I'm a Little Bit of Everything"' in *Sexuality and Culture*, 6, No. 3 (2002), pp. 75–94.

15. See also Alyson Krueger, 'Porn Gets Naughty' in the *New York Times* (28 October 2017), available at: www.nytimes.com/2017/10/28/style/virtual-reality-porn.html

16. Matthew Wood, Gavin Wood and Madeline Balaam, "'They're Just Tixel Pits, Man'": Disputing the "Reality" of Virtual Reality Pornography through the Story Completion Method' in *CHI '17: Proceedings of the 2017 CHI Conference on Human Factors in Computing Systems* (May 2017), pp. 5439–51 (p. 5447).

17. Susanna Paasonen, *Carnal Resonance: Affect and Online Pornography* (Cambridge, MA: MIT Press, 2011), p. 176. See also Gérard Genette, *Narrative Discourse: An Essay in Method*, trans. Jane E. Lewin (Oxford: Basil Blackwell, 1986), pp. 185–8.

18. Zupancic, *What Is Sex?*, p. 56.

19. Jay Owens, 'Post-Authenticity and the Ironic Truths of Meme Culture' in *Post-Memes: Seizing the Memes of Production*, ed. Alfie Bown and Daniel Bristow (London: Punctum, 2019), pp. 77–114 (p. 88). See also Samantha Cole, 'We Are Truly Fucked: Everyone Is Making AI-Generated Fake Porn Now' in *Motherboard*, 24 January 2018, available at: https:// motherboard. vice.com/en_us/article/bjye8a/reddit-fake-porn-app-daisy-ridley

20. McKenzie Wark, 'The Vectorialist Class' in *E-Flux Journal* (May–August 2015).

21. Ibid.

22. For the full argument see Bown, *The Playstation Dreamworld*, pp. 86–8.

23. Ibid., p. 83.

24. Mladen Dolar, *A Voice and Nothing More* (Cambridge, MA: MIT Press, 2010).

25. Barthes, *A Lover's Discourse*, p. 137.

26. Chloe Woida, 'International Pornography on the Internet: Crossing Digital Borders and the Un/disciplined Gaze', a plenary presentation at the *DAC'09 conference on After Media: Embodiment and Context*, University of California Irvine (12–15 December 2009).

27. Kathleen Richardson, 'The Asymmetrical "Relationship": Parallels between Prostitution and the Development of Sex Robots' in *SIGCAS Computers & Society*, 45, No. 3 (September 2015), pp. 290–3.

28. Kate Devlin, *Turned On: Science, Sex and Robots* (London: Bloomsbury, 2018).

29. Isabel Millar, 'Sex-Bots: Are You Thinking What I'm Thinking?' at *Everyday Analysis*, available at: https://everydayanalysis.net/2019/03/15/sex-bots-are-you-thinking-what-im-thinking/

30. See Richard Barbrook and Andy Cameron. 'The Californian Ideology' in *Science as Culture*, 6, No. 1 (January 1996), pp. 44–72.

31. Ibid.

32. Alexander R. Galloway, *Gaming: Essays on Algorithmic Culture* (Minneapolis: University of Minnesota Press, 2006), p. 69.

33. Gilles Deleuze, 'Postscript on the Societies of Control' in *October*, 59 (Winter, 1992), pp. 3–7 (p. 4).

34. Sigmund Freud (1905), *Three Essays on the Theory of Sexuality* in *SE*, 7, pp. 123–245 (p. 168).
35. Illouz, *Why Love Hurts*, pp. 198–9.
36. Ibid., p. 200.
37. Ibid., p. 206.

4 THE MATCH: METAPHOR vs METONYMY

1. See the Alibaba Group description of the early implementation of the score in 2015, available at: www.alibabagroup.com/en/news/article?news=p150128
2. McGowan, *Capitalism and Desire*, p. 178.
3. Sherry Turkle, 'Who Am We', *Wired*, V4.01 (1996), available at: www.wired.com/1996/01/turkle-2/
4. Donna Haraway, 'A Cyborg Manifesto' in *Simians, Cyborgs and Women: The Reinvention of Nature* (New York: Routledge, 1991), pp. 149–81.
5. José Van Dijck, '"You Have One Identity": Performing the Self on Facebook and LinkedIn' in *Media, Culture & Society*, 35, No. 2 (2013), pp.199–215.
6. Julia Ebner, *Going Dark: The Secret Social Lives of Extremists* (London: Bloomsbury, 2020), pp. 52–4.
7. DeAnna Lorraine, *Making Love Great Again: The New Road to Reviving Romance, Navigating Dating, and Winning at Relationships* (USA: 4th Street Media, 2017), p. 152.
8. Georg Simmel, *The Philosophy of Money*, trans. Tom Bottomore and David Frisby (London: Routledge, 2004), p. 73.
9. Ibid., p. 74.
10. Lewis Gordon, 'The Environmental Impact of a PlayStation 4' in *The Verge* (December 2019), available at: www.theverge.com/2019/12/5/20985330/ps4-sony-playstation-environmental-impact-carbon-footprint-manufacturing-25-anniversary
11. Rebekah Valentine, 'Which Gaming Hardware Manufacturers May Have Funded Human Rights Abuses in 2018?' in *Gamesindustry.biz* (June 2019), available at: www.gamesindustry.biz/articles/2019-06-27-which-gaming-hardware-manufacturers-may-have-funded-armed-conflict-in-2018
12. Marijam Didzgalvyte, 'Towards an Ethical Gaming Console' in *Everyday Analysis* (2019), originally published on *Medium* (July 2016).
13. Jessica Ringrose, Laura Harvey, Rosalind Gill and Sonia Livingstone, 'Teen Girls, Sexual Double Standards and "sexting": Gendered Value in Digital Image Exchange' in *Feminist Theory*, 14, No. 3, pp. 305–23 (p. 308).
14. Kishonna L. Gray, *Race, Gender and Deviance in Xbox Live* (Oxford: Elsevier, 2014), pp. 38–9.
15. Ibid., p. 72.
16. Viviana Zelizer, *The Purchase of Intimacy* (Princeton, NJ and Oxford: Princeton University Press), p. 100.
17. Jacques Lacan, *Ecrits*, trans Bruce Fink (London and New York: W.W. Norton, 2005), p. 421.

5 CONCLUSION: READY WORKER ONE

1. Richard Seymour, *The Twittering Machine* (London: The Indigo Press, 2019), p. 3.
2. Ibid., p. 4.
3. Ibid., p. 6.
4. Roisin Kiberd, *The Disconnect: A Personal Journey through the Internet* (London: Serpent's Tail, 2021), p. 231.
5. Charlie Brooker's *How Videogames Changed The World*, 23:45 (4 January 2014), Channel 4, 115 mins.
6. Jacques Lacan, *The Four Fundamentals of Psychoanalysis: The Seminar of Jacques Lacan Book XI*, ed. Jacques-Alain Miller, trans. Alan Sheridan (London: W.W. Norton, 1998), p. 48.
7. Dominic Pettman, *Peak Libido* (London: Polity, 2020), p. 61.
8. Ibid., p. 9.
9. Ibid.
10. Ibid.
11. Mareile Pfennebecker and James A. Smith, *Work Want Work* (London: Zed Books, 2020), pp. 141–2; see also Trebor Scholz, 'How Platform Cooperativism Can Unleash the Network' in *'Ours to Hack and to Own': The Rise of Platform Cooperativism, a New Vision for the Future of Work and a Fairer Internet*, ed. Trebor Scholz and Nathan Schneider (New York and London: Or Books, 2016), pp. 20–6.
12. See Richard Stallman, 'Why "Open Source" Misses the Point of Free Software' in *Communications of the ACM*, 52, No. 6 (2009), p. 31.
13. Christopher Kelty, *Two Bits: The Cultural Significance of Free Software* (Durham, NC and London: Duke University Press, 2008), p. 99.
14. *Free Culture: How Big Media Uses Technology and the Law to Lock Down Culture and Control Creativity*, published under the Creative Commons and available at www.lessig.org, p. 20.

Index

The Pluto Press Newsletter

Hello friend of Pluto!

Want to stay on top of the best radical books we publish?

Then sign up to be the first to hear about our new books, as well as special events, podcasts and videos.

You'll also get 50% off your first order with us when you sign up.

Come and join us!

Go to bit.ly/PlutoNewsletter

Thanks to our Patreon subscriber:

Ciaran Kane

Who has shown generosity and comradeship in support of our publishing.